The Economics of Natural Gas

The Economics of Natural Gas:
Pricing, Planning and Policy

DeAnne Julius and
Afsaneh Mashayekhi

Published by Oxford University Press
for the Oxford Institute for Energy Studies
1990

HD
9560.5
J86
1990

Oxford University Press, Walton Street, Oxford OX2 6DP
Oxford New York Toronto
Delhi Bombay Calcutta Madras Karachi
Petaling Jaya Singapore Hong Kong Tokyo
Nairobi Dar es Salaam Cape Town
Melbourne Auckland
and associated companies in
Berlin Ibadan

Oxford is a trade mark of Oxford University Press

OIES books are distributed in the United
States and Canada by PennWell Books, Tulsa, Oklahoma

© DeAnne Julius and Afsaneh Mashayekhi
1990

All rights reserved. No part of this publication may be reproduced,
stored in a retrieval system, or transmitted, in any form or by any means,
electronic, mechanical, photocopying, recording, or otherwise, without
the prior permission of the Oxford Institute for Energy Studies.

This book is sold subject to the condition that it shall not, by way
of trade or otherwise, be lent, re-sold, hired out or otherwise circulated
without the publisher's prior consent in any form of binding or cover
other than that in which it is published and without a similar condition
including this condition being imposed on the subsequent purchaser.

British Library Cataloguing in Publication Data
Julius, DeAnne
  The economics of natural gas : pricing, planning
  and policy.
  1. Fuel resources: Natural gas
  I. Title II. Mashayekhi, Afsaneh
  III. Oxford Institute for Energy Studies
  333.8233

ISBN 0-19-730011-1

Typeset by Oxford Computer Typesetting
Printed in Great Britain
by Billing and Sons Ltd, Worcester

To our parents

The *Oxford Institute for Energy Studies* was established in 1982 to foster through research and advanced study the mutual understanding of the various parties involved in international energy. The Institute has a publication programme whose aim is to disseminate the results of research undertaken by its staff and associates, selected papers presented at sessions of the Oxford Energy Seminar, and works of high academic quality by outside authors on subjects falling within the research areas of the Institute.

Preface

Like many natural gas projects themselves, this book has been a long time in the making. The idea for it was planted in the early 1980s when we were both economists in the Energy Department of the World Bank. The 1979 oil price increase led to a flood of requests for assistance in natural gas development, and we found ourselves working all over the world on individually complex, but fundamentally similar, problems of gas planning, costing and pricing.

We looked for guidance but found that energy economists had stayed firmly on theoretical terrain, far from the wellhead; while calculations of project costs and benefits – the raw material for planning and pricing analyses – were undertaken by engineers and accountants. The then Director of the Energy Department, Yves Rovani, and his two Assistant Directors, Philippe Bourcier and D. C. Rao, were instrumental in encouraging and guiding us to develop the techniques for bridging the gap between theory and practice.

As experience accumulated, techniques were refined. We benefited greatly in this process from the expertise of project staff elsewhere in the Bank, developing country officials with whom we worked and private consulting companies whose work we helped to supervise. Special mention should be made of Jensen Associates (on LNG), Sofregaz (on residential and commercial distribution), Kennedy & Dunkin (on gas for power) and Chem Systems (on chemical uses of gas). But there was no time during this period to write books.

By the time oil prices fell in the mid-1980s, and many gas projects were put on hold, our ideas were well tested. Although the urgency was less and both of us had moved away from natural gas in our day-to-day work, we were convinced that the time for gas development would return and that future project economists and gas planners should be spared the trials and errors we had been through. We began to write, drawing as much as possible on actual cases, updated to the present.

As luck would have it, gas is once more firmly on the energy scene, partly for oil price reasons but this time also for important, and we believe long-lasting, environmental ones. And both of us are back in the energy business: Julius as Chief Economist at Shell and Mashayekhi as head of the newly-created Natural Gas Unit in the World Bank.

Our hope is that this book will ease the task of analysts and decision-makers concerned with natural gas by providing a comprehensive but practical framework for addressing the critical economic questions. Earlier versions of Chapters 2, 5, 7 and 9 were published as World Bank Energy Department Papers, as cited in the relevant chapters. A variant of Chapter 2 appeared in the Spring 1987 issue of the *OPEC Review*.

We are grateful to the many colleagues and anonymous reviewers who provided comments on the whole and parts of this manuscript. We owe a special debt of gratitude to Robert Mabro, Director of the Oxford Institute for Energy Studies, whose insightful critique and suggestions greatly improved the manuscript, and to his colleague and editor, David Guthrie, without whose diligence and support we might have flagged in the final stages. Any remaining errors are our own.

Finally, our thanks to Ian and Michael who always put up with, and sometimes made possible, those week-ends and evenings in the study that substituted for too many other things while this book was being written.

DeAnne Julius *Afsaneh Mashayekhi*
London *Washington DC*

Contents

	Preface	vii
	Tables	xi
	Figures	xiii
	Abbreviations	xiv
1	Introduction	1

Part I: Background

2	The Analytical Framework	9
	Annex to Chapter 2	20
3	An Overview of Natural Gas Supply and Demand	29

Part II: The Costs of Gas Development

4	The Economic Cost of Gas Development	47
5	Gas Development Costs in Eight Countries	55

Part III: The Benefits from Gas Development

6	Gas Benefit Valuation	65
7	The Value of Gas in LNG Exports	73
8	The Value of Gas in Power and Industry	91
9	The Value of Gas in the Residential and Commercial Market	101

Part IV: Natural Gas Pricing and Planning

10	The Principles of Gas Pricing	115
11	Natural Gas Pricing in Bangladesh: a Case-Study	125
12	A Practical Approach to Gas Planning	145
	Index	173

Tables

A2.1	Estimates and Projections of Gas Demand. 1986–2010.	22
A2.2	Current and Projected Gas Production and Demand from Stage II Field Development. 1986–2009.	23
A2.3	Depletion Date Scenarios. 2010–20.	24
A2.4	Gas Depletion Premiums Assuming Alternative Discount Rates and Rates of Increase of Real Coal Prices. 1990–2015.	26
A2.5	Economic Price of Gas. 1990–2015.	27
3.1	Composition of Selected Raw Gases.	31
3.2	World Proven Reserves of Natural Gas. End-1988.	33
3.3	Flaring as a Percentage of Gross Gas Production in Selected Countries. 1973 and 1988.	35
3.4	World Natural Gas Production. 1988.	36
3.5	World Natural Gas Consumption. 1988.	40
3.6	Structure of World Natural Gas Consumption by Sector. 1986.	42
5.1	Present Value Exploration, Development and Transmission Costs and Projected Volumes of Natural Gas Production in Eight Countries. 1986–2005.	56
5.2	Marginal Costs of Natural Gas at Well-head and City Gate in Eight Countries.	58
7.1	LNG Exports by Country of Origin. 1988.	75
7.2	Evolution of Gas Prices, c.i.f. before Regasification. 1983–8.	77
7.3	Liquefaction Plant Cost Breakdown.	82
7.4	Cost Breakdown at Receiving Terminal under Various Scenarios.	82
7.5	Assumptions for Seven LNG Simulations.	84

7.6	Assumed Price and Unit Netback Values.	86
7.7	Net Present Values of Projects under Different Gas Cost Assumptions for Scenario A.	88
8.1	Average Netback Value of Gas in the Power Sector in Example Country.	94
8.2	Incremental Netback Value of Gas in the Power Sector in Example Country.	94
9.1	Projected Levels of Annual Average Household Consumption for Various Model Distribution Networks.	104
9.2	Average Incremental Costs of Gas Distribution.	109
9.3	Netback Values at the City Gate of Gas Used in the Residential and Commercial Sector.	110
11.1	Incremental Average Day Demand Forecast in Titas System. 1989–2000.	127
11.2	Marginal Cost of Expanding Natural Gas Production at Titas and Habiganj Fields. 1984–2000.	130
11.3	Marginal Costs of Gas Production.	132
11.4	Present Values of Average Incremental Cost of Transmission and of Demand for Titas System.	133
11.5	Customer-related Distribution Cost in the Titas Franchise Area for Different Pipeline Sizes.	135
11.6	Incremental Distribution Costs for Titas System.	137
11.7	Average Incremental Costs for Titas System.	138
11.8	Summary of Marginal Costs of Gas in the Titas System.	139
11.9	Comparison of Existing Gas Tariffs with Economic Costs. June 1986.	141
11.10	Possible Gas Tariff Structure for Titas System.	143
12.1	Sample Presentation of Sensitivity Test Results for Different Cases.	169

Figures

2.1	Volumes and Values of Gas Demand in Various Uses in a Middle-income Developing Country.	11
2.2	Short-run Marginal Cost of Gas.	12
2.3	Long-run Marginal Cost Function for Gas.	13
2.4	Gas Production over Time.	14
2.5	Stepped Marginal Supply and Demand Curves for Gas Without Allowance for Depletion Premium.	16
2.6	Stepped Marginal Supply and Demand Curves for Gas to Determine Depletion Premium.	17
3.1	Shares of Natural Gas Reserves held by Major World Regions. End-1988.	33
3.2	Various Countries' Shares of LNG Imports and Exports. 1988.	36
3.3	Shares of Natural Gas Consumption held by Major World Regions. 1988.	41
7.1	An LNG Export Scheme.	79
7.2	Liquefaction Plant Cost.	80
9.1	Construction Schedule for Network and Service Lines.	106
12.1	Components of a Gas Planning Model.	150
12.2	Demand for Gas by the Power Sector. 1990–2000.	157
12.3	Demand for Gas by the Power and Other Major Sectors. 1990–2000.	158
12.4	Aggregate Demand for Gas. 1990–2000.	159
12.5	Construction of Sample Project Packages.	163

Abbreviations

AIB	Average incremental benefit
AIC	Average incremental cost
bbl	barrels
bcf	billion cubic feet
bcm	billion cubic metres
boe	barrels of oil equivalent
c.i.f.	cost, insurance and freight
CPEs	Centrally planned economies
FAO	Food and Agricultural Organization (UN)
f.o.b.	free on board
GPM	Gas planning model
IGAT	Iranian Gas Trunkline
IGU	International Gas Union
LNG	Liquefied natural gas
LPG	Liquefied petroleum gas
mcf	thousand cubic feet
mcf/h	thousand cubic feet per hour
mmbtu	million British thermal units
mmcf	million cubic feet
mmcf/d	million cubic feet per day
MW	megawatts
NGLs	Natural gas liquids
NPV	Net present value
OECD	Organisation for Economic Co-operation and Development
OPEC	Organization of the Petroleum Exporting Countries
p.a.	per annum
PPNPV	Project package net present value
scf	standard cubic feet
tcf	trillion cubic feet
toe	tonnes of oil equivalent
UAE	United Arab Emirates
UNIDO	United Nations Industrial Development Organization

1 Introduction

Natural gas occupies a curious position in the international energy market. Like oil, it can fulfil a wide range of energy needs from premium feedstocks and environmentally benign space heating to boiler fuel uses. Unlike oil, however, its costs of transport are high so that the major proportion of it is consumed in the country where it is produced. In this respect it is similar to electricity and, to a lesser extent, coal. For the same reason, there is no 'world price' of natural gas, as there is for oil and coal; the market is so thin, and trading patterns so inflexible, that each deal must be tailor-made.[1]

There is an important environmental premium attached to gas which is a cleaner fuel than oil or coal. While this book excludes an estimation of such a premium, none the less energy planners and policy-makers should include the relative environmental costs and benefits of gas and alternative fuels in project appraisal.

Under such conditions, it is not surprising that natural gas prices vary widely around the globe or that one country will be heavily dependent on gas for its electricity generation while a neighbouring country will limit its use of gas to fertilizer and petrochemical production. Such seeming anomalies result from the mixture of technical and market forces that make the economics of natural gas such a varied and interesting subject.

It is also a subject that is neither well understood nor widely pursued. In the industrialized countries, despite the existence of long-established and mature gas markets, the techniques of regulation and pricing are mostly relics of historical accident,

---

[1] This situation is changing in the United States where extensive pipeline networks and a unique ownership pattern have made possible a deregulated gas industry with the numerous buyers and sellers necessary to produce a competitive market-place. In most countries, however, natural gas is still a vertically integrated business which is run either as a public monopoly or as a regulated private one.

based on the inappropriate application of accounting principles to quintessentially economic concerns.

In the developing countries, the concerns themselves vary as widely as do the attempted remedies. In some, the problem starts at the well-head. Systems of producer pricing and taxation have been borrowed from the oil industry but, unless modified to suit the local gas situation, they will seriously hamstring gas development. In other countries, the main problem is consumer prices. Gas must be competitive with a whole range of other fuels in different uses, and this often tempts governments to try to 'cream the market' by applying a complex, use-specific array of gas tariffs. Unless these are perfectly calculated, and adjusted every time other energy prices or demand factors change, they provide little incentive and much uncertainty for potential consumers who are considering a switch to gas. In many developing countries, not only gas prices but the full range of gas development questions need to be addressed. These include the allocation of gas among competing uses, the sequence of field development, the reduction of flaring, and the speed of depletion, to name but a few. Traditional rules of thumb from engineering texts will not suffice: to answer such questions requires a comprehensive framework of analysis that can be sensibly applied to local technical and economic circumstances.

1.1 The Obstacles to Gas Development

Natural gas development faces a number of obstacles that have heretofore slowed the market penetration of gas, especially in developing countries. Reviewing the obstacles is a useful way to preview the contents of this book.

The first obstacle to rapid gas development is its *structure of costs*. Large, up-front investments are needed in pipelines and distribution systems before the gas can begin to be used. In this respect, gas is similar to electric power, water supply and other public utilities. It is difficult to start with small investments, even though initial demand will be small. This creates financial problems for the producer just at the time he is needing to expand his investments to meet future demand.

In addition to the financial problems inherent in infrastructural development, economic uncertainties arising from the structure of gas costs have sometimes delayed projects. There has been no single standard in the gas industry for how gas costs should be calculated in order to make easy and valid comparisons either with the costs of alternative energy investments or with the benefits of the gas investment. In Part II we examine the various possible measures of costs that have been used for other types of projects with 'lumpy' investments. Then we apply a consistent standard to compute the costs of gas development in eight countries.

A second obstacle to rapid gas development is the *misperceptions about the 'best' use of gas* that policy-makers, planners, engineers and economists have perpetuated. Perhaps because of the role that gas now plays in most industrial countries – a premium fuel used principally as an industrial feedstock and for residential heating and cooking – there is a pervasive bias against using gas as a boiler fuel. This bias ignores the widely differing supply and demand conditions under which gas occurs in developing countries, as well as the critical role that interruptible large consumers (which are nearly always fuel burning) can play in the early stages of infrastructure utilization. In addition, the economics of both the feedstock and fuel uses of gas in developing countries differ significantly from those in industrial countries: it is not unusual to find fuel uses yielding higher benefits than feedstock uses. In Part III we examine the various ways that gas benefits can be measured, and calculate the economic benefits from gas in its major uses.

The third important obstacle to gas development is the widespread *inappropriateness of domestic gas prices*. The regulated structures for gas pricing in most industrial countries provide few useful models. Gas prices in developing countries often reflect the outcomes of ancient negotiations between individual producers (usually of associated gas) and consumers. In the absence of useful cross-country or historical guidelines, one might turn to the theory. But here also there are problems. There are disagreements among economists about issues such as the relevance to gas pricing of utility pricing models and depletable resource pricing models, whether gas is a tradable good with a border price, and which is the best tax structure to capture the resource rent for the government. In

Part IV we review the relevant theoretical basis for gas pricing and develop a detailed case-study of tariff calculation.

A final obstacle to gas development arises because of the geological link between gas and oil. It is still the case that most gas is discovered in the course of drilling for oil. In the oil business, of course, non-associated gas discoveries are regarded as only slightly better than dry holes. Thus it is hardly surprising that *leaving the responsibility for gas development with the gas discoverer* results in progress at a snail's pace.

Managing gas development during its early stages is not easy. It often seems a 'chicken-and-egg' situation: without the infrastructure there is no gas to use, but it is hard to justify the large investment in infrastructure when there is little historical use from which to project demand. In addition, because of the depletable nature of gas and the complementarities of various uses, decisions on individual gas-producing or gas-using projects cannot be made independently. Trade-offs over time, as well as trade-offs between alternative uses at any point in time, must be explicitly addressed at the pre-investment stage.

This kind of planning cannot be done on the back of the proverbial envelope. Again, the analogy with electric power is relevant. Just as a long-term power system development model provides the framework within which individual project alternatives can be properly compared, so in natural gas we advocate the use of a formal, long-term, sector model in project selection and design. The next chapter presents an analytical framework for gas planning, and the final chapter translates that framework into a practical planning tool.

1.2 The Approach of this Book

This book is an attempt to break away from stereotyped thinking about, and rules of thumb for, gas development. Except under unusual conditions, gas is neither a 'noble' fuel nor a bothersome by-product of oil production. Its value, and thus its appropriate price and use, will depend on local demand and supply conditions. These can be analysed in a systematic way to address issues of pricing and planning.

Most of our examples are drawn from developing countries,

and most of our analysis starts from first principles rather than focusing on an industrial structure and regulatory system already in place. For those readers involved in the development of a new gas market, such an approach is obviously suited. We believe it is also a useful way to shed new light on many of the key issues facing a mature gas industry. Techniques of comparing the costs of alternative field developments, of reconciling multiple objectives in tariff design, and of evaluating different gas usage patterns are as relevant to established gas utilities as to new ones.

The objective of this book is to draw together in one volume the key concepts involved in the economics of natural gas and to illustrate their practical application. In so doing we seek both to present an accurate perspective on the costs and benefits of gas development and to suggest policies and practices – particularly for pricing and planning – that would enhance the prospects of rapid gas development and economic gas use.

While our range of topics is a broad one, our perspective is primarily economic. Thus, except in so far as they relate to gas costing, we do not attempt to cover the financial or accounting practices of gas utilities. Nor do we deal with the regulatory mechanisms that have been used to influence tariffs or to control rates of return. While these are important aspects of gas development, they undergo continual change and are outside our purview.

As far as possible, the approach of the book is inductive, and therefore empirical. Our cost and benefit figures are based on actual projects or gas sector analyses carried out in a number of developing countries during the 1980s. Much of this work was initiated through World Bank projects or studies, but we have also used studies and analyses carried out by developing country officials, consulting firms and others.

In addition to the empirical analyses, each part of the book begins with a chapter outlining the theoretical background for that area. There are many grey areas and some important controversies in the literature over definitions and over the approach to be used in costing, valuing benefits, pricing and planning for gas. These chapters have been written for the educated layman, with a minimum of jargon and mathematical notation. However, at some loss of depth, they can be skipped if the reader is interested only in the empirical results.

These results indicate to us that, even with the lower oil prices of the late 1980s, there are many countries that could benefit from the acceleration of natural gas development and use. As shown in Part II, in most cases the fully allocated costs of gas supply remain below $10 per barrel of oil equivalent. We also conclude that the size of undeveloped gas reserves is *not* an obstacle to greater gas use. The reserves and the technology exist and are well proven. The key issues are economic ones:

(a) How much does it cost to develop?
(b) What is it worth in the market-place?
(c) How should prices be set to provide incentives to potential producers and users while capturing a fair resource rent for the government?
(d) Can gas development be planned in a systematic and practical way?

It is at such questions that this book is addressed.

PART I

BACKGROUND

2 The Analytical Framework

As noted in Chapter 1, for most countries with an evolving natural gas industry, issues relating to the costs, benefits, prices and planning of gas must be addressed. While dealing with these issues individually, however, it is important to recognize their inherent linkages. This can be done by presenting, at the outset, a general framework for gas analysis, through which the various pieces can be seen in context. To keep the focus in this chapter at a general level, we present below a simple framework which subsumes many of the detailed issues that are taken up in later chapters.

The framework is built around an analysis of current and projected gas demand and supply over a medium- to long-term period. Using this information, it is possible to estimate the economic price of gas over that period. The economic price – or opportunity cost – of gas is a key parameter for evaluating individual gas development projects, for gas allocation choices, as a pricing or negotiating bench-mark, as a guide for policies on flaring or depletion, and for many other types of decisions regarding gas development and use. Of course, the planner is still faced with uncertainties regarding the future prices of competing energy products, the growth of world markets for 'embedded' gas exports such as urea, and so forth. However, such uncertainties beset most long-term investment decisions. The special problem for the gas analyst is how to determine the value of the gas itself.

Why is this difficult? As discussed below, in some countries it is not. Where gas reserves are small and substitute at the margin for some tradable product such as fuel oil or coal, then the economic price P_e of gas is clearly the equivalent cost of the tradable product (suitably adjusted for thermal and other technical differentials). At the other extreme, where gas reserves are very large relative to prospects of increased demand, both domestic and for export, then P_e will be close to the production cost of the gas. As a first guidepost, it is sometimes useful to remember that the economic price of gas will always be between the two values described by these two cases. The problem arises when: (a) there is a large difference between the cost of production and the price of the marginal fuel replaced; and (b) gas reserves are neither very large nor very small relative to potential demand. It is really for these intermediate cases that the full framework described in this chapter is needed. In the final section we discuss how the general framework can be used to categorize countries (e.g. gas-surplus, gas-short) in order to identify analytical short cuts and to highlight, at an early stage, issues of particular importance to the development of their gas sectors.

2.1 The Pattern of Gas Demand

Natural gas is a flexible fuel with many potential uses. Its highest-value uses are often where it substitutes for expensive petroleum products such as diesel fuel (e.g. for peak power generation) or kerosene (e.g. for household cooking). In most countries, however, the small quantities of gas that would be required to satisfy such uses fully would not be sufficient to justify the high initial costs of gas development and transmission. Larger-volume – but lower-value – uses of gas are as replacements for fuel oil or coal in the power and industrial sectors. Small amounts of gas can be used as a feedstock for fertilizer or methanol production; the value of gas in these uses will vary with the world prices of such products. Recently, these prices have been low because of global over-capacity, including that of gas-based export plants in the OPEC countries. Thus, while the relative value of – or willingness to pay for – gas in its different uses will vary both over time and across countries, the demand curve for gas will generally consist of a series of steps, such as that shown in Figure 2.1.

The height of each step represents the value of the gas as derived from the export or import prices of the goods it is used to produce. The highest-value uses of gas are for peak power generation (where it essentially replaces diesel oil), household distribution (where it replaces LPG and kerosene) and metha-

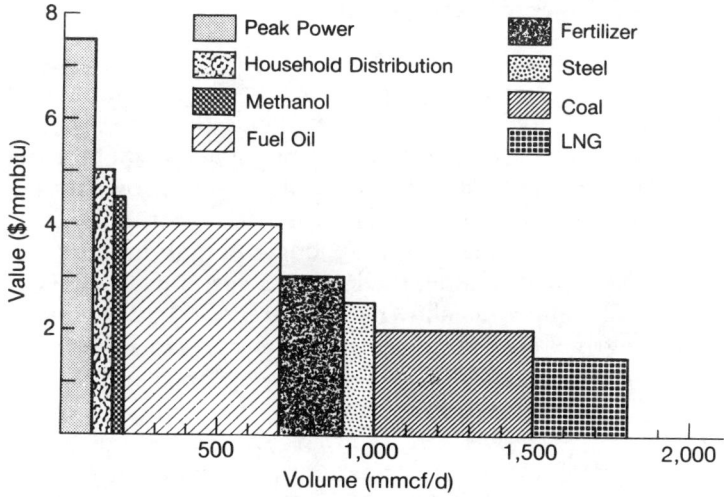

Figure 2.1: Volumes and Values of Gas Demand in Various Uses in a Middle-income Developing Country.

nol. The total amount of gas that can be consumed for these purposes, however, represents only a small share of the potential domestic market, leaving aside for now the export market for LNG. The replacement of fuel oil and coal in the power and industrial sectors clearly constitutes the bulk of the market. The fertilizer and steel plants based on gas are sized to replace all imports of those commodities and generate a surplus for potential export until domestic demand for these commodities will be large enough to absorb the full output. The value and size of LNG export are based on a facility sized to produce the maximum amount the country could reasonably expect to sell. For these reasons, if one were to visualize how the curve in Figure 2.1 would shift over time, certain steps would grow longer (i.e. those based on domestic demand such as household distribution and fuel oil replacement) while others (i.e. fertilizer production and LNG export) would probably remain unchanged.

2.2 The Pattern of Gas Supply

Leaving aside for the moment the fact that total gas supply is constrained by fixed total reserves, the supply functions for gas look much like those for other goods that are subject to economies of scale. Investment in new capacity will take place in lumps, with periods in between of relatively low short-run marginal costs (see Figure 2.2). It is often desirable to calculate a smoothed estimate of these long-run marginal costs (see Chapter 4). If this is done, the long-run cost function for gas is likely to be stepped as shown in Figure 2.3.

The length of each step in Figure 2.3 represents the amount of sustained production that could be developed and delivered for the incremental cost plotted on the vertical axis. For example, the first (lowest) step shows an amount of 200 mmcf/d of onshore, associated gas available at an incremental cost of $0.20/mmbtu. The second and third steps show production of non-associated gas from onshore and offshore fields, respectively, at progressively higher costs. The sum of these three steps gives the country's potential gas supply based on its estimated reserves.

The technical characteristics of gas supply mean that over time production will follow a path such as that illustrated in

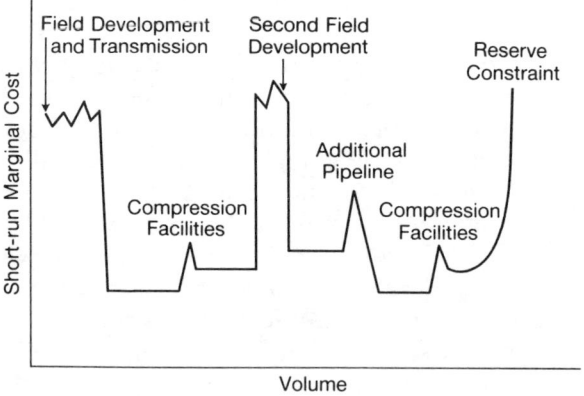

Figure 2.2: Short-run Marginal Cost of Gas.

Figure 2.4. Potential production will rise in increments as new fields are discovered and developed and as new transmission pipelines are laid. At some point, shown as T^* in Figure 2.4, however, the reserve constraint will become operative and production will plateau. The productive life of the last fields developed will determine the length of the plateau. For both technical and economic reasons (related to the life of gas-using facilities and thus the length of supply contracts), it is often 10–15 years. After that production falls off rapidly as the remaining fields are exhausted (at T^e in Figure 2.4).

2.3 The Economic Price of Gas

The economic price of any good – gas included – is determined by the intersection of its aggregate demand and supply curves. For many goods, that value is simply the price resulting in an open market-place through the bargaining of many buyers and sellers. For goods that are traded internationally, such as oil or wheat, the economic price in a small country is often determined almost independently of the local demand and supply conditions.

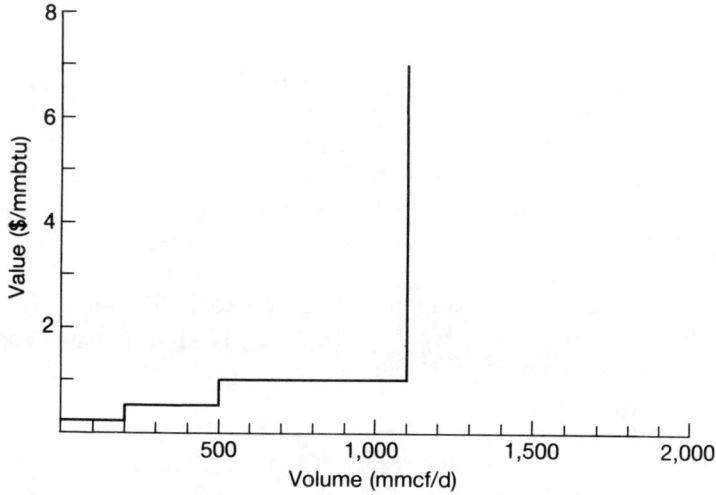

Figure 2.3: Long-run Marginal Cost Function for Gas.

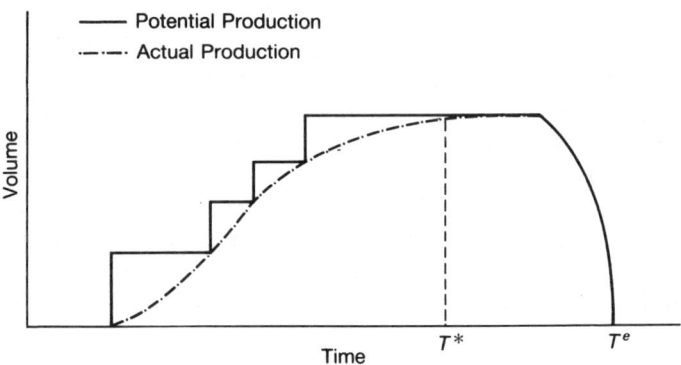

Figure 2.4: Gas Production over Time.

Unfortunately, there are several reasons why the economic price of gas generally cannot be so easily observed. First, it is not a widely traded commodity and, owing to the high transport costs involved as well as the presence of competing imports by pipeline to the major consuming markets, the export of LNG is not an economically viable option for most countries. Even those that do export face clear market constraints to the expansion of their exports so that, at the margin, the value of their gas cannot be derived directly from its export price.

Second, although oil is a traded good, gas is not a perfect substitute for all petroleum products. In some uses, such as home heating and electricity generation, the two are closely substitutable; but for most transport uses, gas is only an oil substitute at a prohibitively high cost. As a feedstock for petrochemical and fertilizer production, oil products are a poor substitute for gas; and the alternative to domestic production of fertilizer based on gas is to import it directly rather than to produce it using oil as a feedstock. Because of this imperfect substitutability, the fuel oil-equivalent price of gas is not always an appropriate short-cut estimate for the economic price of gas.

Third, in domestic markets for gas, political and economic factors have generally led to price regulation because of monopolistic production and distribution. In many developing countries, the demand side is also characterized by an effective monopsony held by the electricity company, which often accounts for more than three-quarters of purchases. Thus, whatever price bargaining takes place is often between two large state enterprises, rather than many small buyers and sellers. If unregulated, the outcome will reflect their relative bargaining skills and political clout rather than the underlying economic parameters.

The final reason that the economic price of gas may not be immediately observable arises because its total stock is fixed. Unlike wheat or widgets, gas is a depletable commodity. A country has a fixed (though generally unknown) stock, and consumption of a btu today means forgoing the consumption of a btu at some future date. The value of this forgone consumption has been called many things in the economic literature: depletion premium, royalty, user cost, net price or resource rent. We use the term 'depletion premium' to refer to the *present value* of the forgone consumption. The key pricing distinction between a depletable and a renewable resource is that the opportunity cost of the former will include a depletion premium as well as its production (or extraction) cost.

Returning to the supply and demand curves described above, if gas were not a depletable resource, then the two curves could simply be superimposed to determine the economic price. It is useful to follow this step through for illustrative purposes to show how – with stepped demand and supply curves – the economic price could be represented by either a marginal willingness to pay or a marginal supply cost, depending on which constraint is binding. This is shown in Figure 2.5.

In the period shown in Figure 2.5, production from proved reserves would be sufficient to meet all of the uses down to and including steel production and about one-fifth of the coal substitution. This would indicate that the economic price of gas would be its value as a coal replacement, equivalent to about $2.00/mmbtu. If, however, the marginal cost of the last block of gas were $2.20/mmbtu, as shown by the dashed line, the intersection of the supply and demand curves would be on a horizontal segment of the supply curve. This means that the opportunity cost of gas would be equivalent to its incremental supply cost, and no coal substitution would take place.

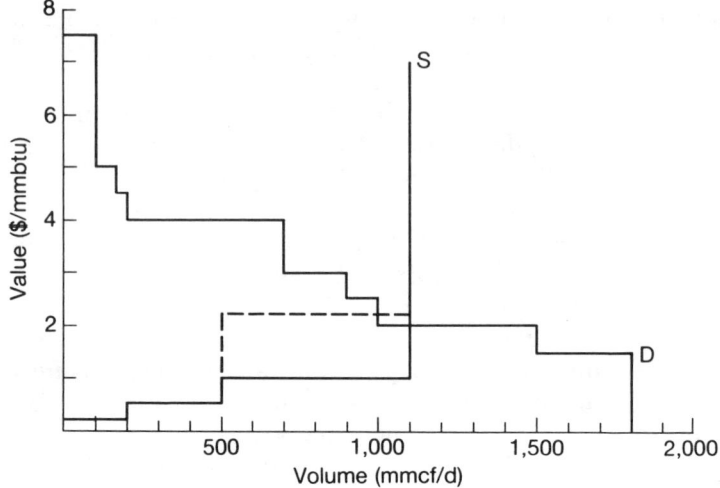

Figure 2.5: Stepped Marginal Supply and Demand Curves for Gas Without Allowance for Depletion Premium.

The above analysis, in which we developed and superimposed the demand and supply curves to find their intersection, would be the appropriate technique for determining the economic price of a non-depletable resource. However, for a depletable resource, there is an additional consideration. As we saw above, the prices and production decisions in different time-periods are linked for such commodities. Therefore the economic price today must include an element – the depletion premium – that represents the forgone opportunity of consuming the resource in the future.

The depletion premium concept can be illustrated by looking again at our example. Suppose that in five years' time the demand curve – now labelled D_0 in Figure 2.6 – shifts out to D_1. The marginal use of gas at that time is for fertilizer production, and its value is $3.00/mmbtu. Because the gas is depletable – i.e. the total stock is fixed – the question arises whether it is better to use that marginal unit of gas today and save $2.00/mmbtu on coal imports or to hold it for five years and use it then for fertilizer production where its value is $3.00/mmbtu.

Suppose the owners of the gas have the option of investing their profits at a 10 per cent rate of interest. The profit per unit

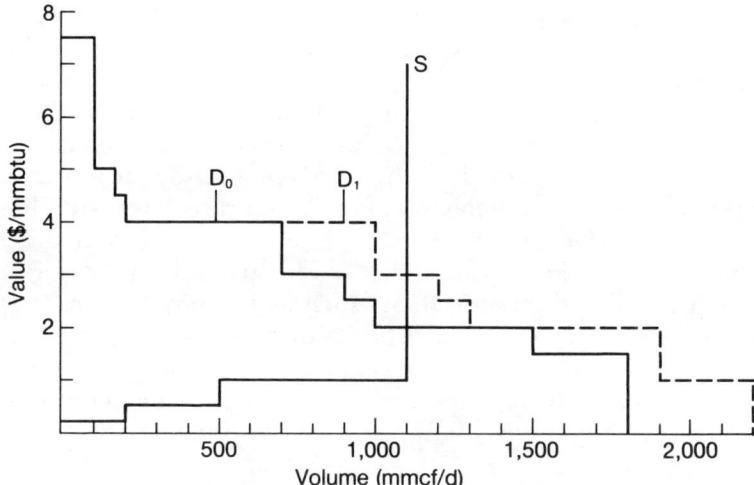

Figure 2.6: Stepped Marginal Supply and Demand Curves for Gas to Determine Depletion Premium.

of gas sold today would be its price, $2.00, less its production cost (from the supply curve) of $1.00, i.e. $1.00. If this $1.00 were invested at a 10 per cent interest rate, its value after five years would be $1.61. However, if the gas were left in the ground to be exploited later for fertilizer production, the owners would earn a profit of $2.00 (i.e. the price of $3.00 less the production cost of $1.00). Since this is greater than the future value of today's profit, $1.61, the gas should be left in the ground. Put another way, the lowest price that the owner should accept for the gas today is $2.24/mmbtu – i.e. its production cost of $1.00 plus the present value of the forgone future profit, $1.24.[1]

This example illustrates the basic principle of depletable resource economics, as originally developed by Harold Hotelling in 1931.[2] The economic price of a depletable resource has

---

[1] $1.24 is the present value, at a 10 per cent discount rate, of $2.00 in five years' time.

[2] Hotelling, H., 'The Economics of Exhaustible Resources', *Journal of Political Economy*, Vol. 39 (April 1931), pp. 137–75. For a non-technical development of this result, see Webb, M. G. and Ricketts, M. J., *The Economics of Energy*, Macmillan, London, 1980, Chapter 3.

two additive elements – its extraction cost and a depletion premium which rises over time at the discount rate. The Hotelling theory is strictly valid only under a set of rather restrictive conditions. These include the existence of perfectly competitive markets for both current and future goods and certainty as to the stock of the resource and the current and future shape of the demand curve. Much recent research has gone into exploring the results of relaxing these and other conditions on the original theory. The general thrust of it is that with diverse assumptions, such as technological change reducing the extraction costs or the discovery of new reserves, it is possible to cause the market price of the resource to move along almost any time-path, even a declining one. The basic results remain robust, however, over a wide set of assumptions.

2.4 Application of the Framework to Country Classification

As mentioned in the introduction to this chapter, there are many situations in gas planning and pricing where the calculation of a depletion premium will not be necessary or where other short cuts to the full analysis presented above can be taken. Using the general framework, we define three classes of countries – gas-surplus, gas-short and surplus-window – and discuss below the aspects of the general framework likely to be of greatest importance to each. This provides a preview of the kinds of issues taken up in Part IV.

A 'gas-surplus' country is one in which the demand–supply balance is such that T^*, the point of economic depletion, is very far into the future. Once T^* is 40–50 years away, the present value of the depletion premium at T^* is insignificant when compared with the ranges of uncertainty surrounding other estimates such as the cost of gas development. Thus, for practical purposes, the depletion premium can be ignored. For such a country, questions of project sequence will not be paramount, since gas availability is not a constraint. The constraints to bringing projects on stream more rapidly will often be economy-wide ones: the availability of capital, managerial skills and industrial infrastructure. Therefore careful attention should be paid to exploiting potential complementarities and economies of scale in gas investment.

A 'gas-short' country is one in which the potential availability of gas is projected never (or only briefly) to exceed the potential demand for it. In a number of countries, current fuel oil consumption is several times as large as natural gas production. Even if the latter increases significantly, with projections of continued economic growth, it is likely that incremental gas supplies will continue to replace fuel oil at the margin. In such a case, the economic price path for gas will follow that of fuel oil, so long as the cost of incremental gas supplies remains below that level. Since consumption will be supply constrained in a gas-short country, more time should be spent on the demand side of the analysis than on the supply side. The cost of gas development can be calculated roughly, simply to ensure that its costs remain below the cost of the replacement fuel. Demand analysis should focus on identifying specific project candidates in the main domestic markets (e.g. electricity, fertilizer, cement), rather than on estimating long-run trends in aggregate demand. Project packaging, ranking and selection will be the most critical part of the analysis for a gas-short country. With gas availability a constraint over the entire period, questions of project size and sequence are of major importance.

A 'surplus-window' country is one for which the planning period includes times of both gas surplus and gas shortage. T^* is either projected to fall within the planning period or clearly foreseeable beyond it. In such a case, there are few short cuts to the full analysis that can be taken. Both demand and supply must be estimated carefully in order to derive the price path of gas. A general estimate of the cost of gas development will be needed to ensure that the price path does not fall below it. Issues of project timing will be critical since some potential projects will be viable only during the period of gas surplus, and these should be compared initially on a mutually exclusive basis to find the best candidates for that possibly brief period. Because of the sensitivity of the gas price path to project selection in surplus-window cases, several complete iterations will often be required to arrive at a consistent plan for gas development. An example of this case is given in the annex that follows.

Annex to Chapter 2

The purpose of this annex is to illustrate the calculation of the depletion premium and the economic price of gas in what we have called the 'surplus-window' case. This is where gas reserves are neither very large relative to demand, nor very small. In the former case, the economic price of gas P_e will be approximated by its marginal cost because the depletion premium element will be so small as to be insignificant. For countries with small reserves relative to demand, the economic price of gas will be equal to the price of its marginal replacement fuel, so no complex calculation of the depletion premium is required. To illustrate the depletion premium calculation, we must choose an intermediate case.

The data we use are those of an actual developing country which has recently been developing its natural gas sector. It is well endowed with natural gas but at an early stage of its utilization. We assume that it has some hydro resources but no indigenous coal. The power sector is expanding rapidly as the connection rate is increased, and inherited generating plant is based on fuel oil, hydro and diesel oil (for peaking). Industry also uses fuel oil and diesel for steam raising and on-site power generation. The first step in determining the economic price of gas is to make a rough estimate of demand and supply trends.

A2.1 Demand Estimates

Once gas becomes available, consumption can be expected to increase rapidly as existing industries and power stations are converted to gas. Thereafter, gas consumption can be expected to grow in line with overall energy demand. The country's large gas reserves are sufficient to provide a level of production that far exceeds its current total fuel oil and diesel consumption. If P_e were assumed to be just below the price of fuel oil – the current replacement fuel at the margin – the potential gas supply would far exceed demand, at least in the medium term. Moreover, at that sort of price, the power sector would base its future expansion on imported coal rather than gas. This would leave only industrial users in the gas market, resulting in an even larger surplus as newer coal plant displaced the converted gas plant over time. Importing large quantities of coal while leaving gas in the ground is unlikely to be the optimum energy path for the economy, however, unless the gas development costs are exceptionally high. Therefore, for the initial gas demand estimates we assume that P_e is no higher than the price path of imported coal for use by the power sector. At this stage we do not need to make an estimate of that price.

The first four columns of Table A2.1 show our initial estimates and projections of gas demand in the power and industrial sectors, the export market, and the totals for all three combined. Gas for the export market is exported via a pipeline to a neighbouring country for use in its power and industrial sectors. In order to achieve the consumption levels forecast for the post-1990 period, it will be necessary to build an onshore pipeline to serve the interior of the producing country where the major population centres are located. It is estimated that 1990 is the earliest that such a project could be commissioned. An additional spur would be built to the importing country by 1995 and, once established, the contracted export volumes would double within five years. The final column of Table A2.1 adjusts these 'dry gas' demand figures to 'raw gas' equivalents; the difference being the calorific value of the propane and butane removed from the gas before transmission and the losses incurred through using gas in the compression stations along the pipeline.

A2.2 Medium-term Supply and Cost Implications

Meeting the demand projections shown in Table A2.1 will require a major build-up of gas production and transmission infrastructure. The existing fields and infrastructure can deliver a maximum of 400 mmcf/d, which would be insufficient to meet the 1990 demand. A phased Stage II programme of

Table A2.1: Estimates and Projections of Gas Demand. 1986–2010. Million Cubic Feet per Day.

	Power	Industry	Export	Total Dry Gas	Through Domestic Pipeline	Total Raw Gas
1986	70	10	–	80	–	100
1989	115	15	–	130	–	165
1990	340	38	–	378	241	480
1995	523	144	150	817	650	1,034
2000	767	207	300	1,274	1,107	1,613
2005	850	315	300	1,465	1,268	1,854
2010	1,160	379	300	1,839	1,642	2,328

Note: 1986 data are actual; data for 1989 onwards are estimates.

offshore field development is planned, along with a four-year construction programme for the pipeline. The initial increments of gas production for Stage II will be from large, non-associated fields located close off shore. Production from these fields will plateau at about 1,600 mmcf/d, which will be sufficient to meet demand until about 2005. Beyond that time it is likely that costs will be higher as smaller fields are developed.

Table A2.2 shows the information needed to calculate the cost of Stage II field development.[1] With a discount rate of 10 per cent the average incremental cost for gas production is $1.03 per mcf of raw gas. We assume that the cost of the next development sequence – Stage III – would be about 17 per cent higher, or $1.20/mcf.

Against these costs must be offset the value of the liquids that are produced as a by-product with the gas and separated at the gas processing plant for independent sale. Current raw gas production in the producing country contains 9 per cent LPGs and 11 per cent condensates. Together these would be worth about $0.60/mcf. Deducting the cost of the processing plant – about $0.20/mcf – leaves an offset benefit of $0.40/mcf. If oil prices rise, this benefit will increase. On the other hand,

---

[1] See Chapter 4 for a discussion of this method and the assumptions behind it.

the liquid component of the fields developed in Stage II may be lower than that of currently producing fields, or processing costs may be higher if carbon dioxide or other impurities are present in the gas. Given these uncertainties, we assume the offset benefit from the liquids to be constant over the projection period.

Table A2.2: Current and Projected Gas Production and Demand from Stage II Field Development. 1986–2009. Volumes in Million Cubic Feet per Day. Costs in Million 1986 Dollars.

	Demand	*Production*	*Total Capital and Operating Costs*
1986	100	–	8
1987	125	–	20
1988	145	–	97
1989	165	–	312
1990	480	80	43
1991	560	160	30
1992	654	252	99
1993	761	361	240
1994	887	487	128
1995	1,034	634	290
1996	1,130	730	85
1997	1,235	835	67
1998	1,350	950	178
1999	1,476	1,076	466
2000	1,613	1,213	312
2001	1,700	1,300	310
2002	1,791	1,391	299
2003	1,887	1,487	462
2004	1,861	1,461	592
2005	1,854	1,454	816
2006	1,942	1,542	485
2007	2,047	1,600 (maximum)	613
2008	2,158	1,600	739
2009	2,274	1,600	279
NPV at 10 per cent:		4,877 (= 1,780 bcf)	1,828
Average Stage II Cost:		1,828/1,780 = $1.03/mcf	

Note: 1986 data are actual; data for 1987 onwards are estimates.

24 The Economics of Natural Gas

The cost of gas transmission facilities must also be estimated. Plans are only available for the initial set of investments which will provide for a delivered capacity of 850 mmcf/d. Using the throughput figures from Table A2.1 and the same methodology as shown for development costs in Table A2.2 yields an average incremental cost of transmission of $0.30 per mcf of dry gas. This figure includes all costs needed to reach the design capacity of 850 mmcf/d. According to our demand estimates, that would happen in 1998. Additional investments in compression or other means of increasing throughput would then be required. However, it is unlikely that their average cost would exceed that of laying the original pipeline. Thus we assume a constant transmission cost over the period.[2]

A2.3 Implications for Depletion

Extrapolating the power and industrial gas demand separately beyond the estimates shown in Table A2.1 yields an average annual growth rate of total demand of just under 6 per cent. Assuming that a fifteen-year production plateau is required to accommodate long-term gas users, Table A2.3 shows the implied levels of plateau production and required reserves corresponding to the dates 2010, 2015 and 2020.

Table A2.3: Depletion Date Scenarios. 2010–20.

	Cumulative Production (tcf)	Additional Reserves for R/P Ratio of 15 Years (tcf)	Plateau Production (mmcf/d)	Implied Total Reserves (tcf)
2010	11.5	13.7	2,500	25.2
2015	16.6	16.6	3,200	32.8
2020	23.4	21.9	4,000	45.3

[2] For tariff purposes the transmission cost should be allocated only to peak users, as discussed in Part IV.

The proven recoverable reserves of the country are 30 tcf. It has been estimated that a further 10–15 tcf may be discovered as exploration proceeds. The first line of Table A2.3 indicates that current proven reserves would be more than sufficient to sustain a production plateau of 2,500 mmcf/d for fifteen years past 2010, when approximately that level of demand would be reached (since 25.2 < 30). If the future discoveries of the anticipated higher level of 15 tcf were realized, then the reserve constraint would not become binding until about 2020, at a plateau production level of 4,000 mmcf/d (sustainable for a further fifteen years). The mid-point of 2015 would represent a modest level of reserve additions (2.8 tcf beyond current proven reserves), and result in a plateau production level of 3,200 mmcf/d. This seems a reasonable base case.

To calculate the depletion premium, we need to know the value of the marginal substitute for gas in the year 2015. On the basis of current projections of fuel prices, the marginal substitute in the power sector is likely to be imported coal. At an average annual growth rate of 1 per cent in real coal prices, the 2015 c.i.f. price of coal would be $60/ton (in constant prices). Taking into account differential capital costs and fuel efficiencies, this is equivalent to $2.60 per mcf of dry gas. The difference between that value and the production and transmission cost of gas in 2015 ($2.60 − $1.20 + $0.40 − $0.30 = $1.50/mcf of dry gas) represents the resource rent accruing to the economy from the use of gas.[3] The present value of that difference is the opportunity cost of using gas at an earlier point which, at the margin, displaces imported coal in the future.

Table A2.4 shows the resulting depletion premiums, under a variety of assumptions. An alternative scenario is postulated where the coal price would increase at 2.5 per cent per annum in real terms. This results in a 2015 price of $92/ton, which is equivalent to $3.80 per mcf of dry gas after adjustments for differential capital costs and fuel efficiencies. It is clear that

---

[3] After 2015, as the power sector shifts out of gas, its economic price will continue to rise in line with its replacement, at the margin, for higher-value uses in other sectors.

Table A2.4: Gas Depletion Premiums Assuming Alternative Discount Rates and Rates of Increase of Real Coal Prices. 1990–2015. 1986 Dollars per Thousand Cubic Feet of Dry Gas.

Coal Price Increase:	1.0 % p.a.		2.5 % p.a.	
Discount Rate:	5%	10%	5%	10%
1990	0.4	0.1	0.8	0.2
1995	0.6	0.2	1.0	0.4
2000	0.7	0.4	1.3	0.6
2005	0.9	0.6	1.7	1.0
2010	1.2	0.9	2.1	1.7
2015	1.5	1.5	2.7	2.7

the choice of discount rate has a greater effect on the depletion premium – and thus on P_e – than does the choice of assumption about the future path of coal prices. We also note that the discovery of an additional 15 tcf of gas – 50 per cent more than today's proven reserves – would reduce the depletion premium by a mere $0.10/mcf.

If we assume that coal prices do not increase over the period 1990–2015, then the value of the depletion premium in 1990 would be $0.20/mcf rather than the $0.40/mcf shown in the first column of Table A2.4, and the depletion premium is still about one-third the size of the net production cost of the gas, $0.60/mcf. In order for the depletion premium to be zero, coal prices would have to fall from $45/ton to $25/ton at some point in the future.

The choice of discount rate for the calculation of the depletion premium is a difficult issue. Many feel that the rate used for project comparison and selection is too high. That rate presumes to incorporate both the marginal productivity of investments and the time value of present versus future consumption. It is appropriate for relating paired streams of benefits and costs. For gas pricing, the primary concern is the trade-off between present and future income. Moreover, in project work the weighting of distant future benefits and costs is usually less important, and there is little concern over whether the appropriate rate is constant or falling over time.

A lower discount rate has the effect of increasing the current cost, which may depress consumption below its optimum, but which is also advocated as the risk-minimizing stance for such irreversible decisions. As noted earlier, the range of uncertainty that applies to calculations of the depletion premium is large, which also argues for incorporating a margin for error in the price. The lower rate (i.e. the higher depletion premium) is taken as our base case for these reasons and as a quantitative reflection of the country's preference for resource conservation.

A2.4 Summary of the Economic Price of Gas

We are now ready to combine the different estimates to arrive at the economic price of gas for this country. To recap briefly, this is the production cost of raw gas, less the net benefit derived from liquid separation, plus the transmission cost (if applicable, depending on the location of the user) plus the depletion premium. These components are summarized in Table A2.5.

In a gas-rich country, such as that of our example, it is necessary to carry the calculation relatively far into the future in order to solve for the near-term depletion premium. It is interesting to note that even with the reserve constraint not binding until the year 2015, 30 per cent of the economic price of gas in 1990 is composed of the depletion premium. It is

Table A2.5: Economic Price of Gas. 1990–2015. 1986 Dollars per Thousand Cubic Feet.

	1990	1995	2000	2005	2010	2015
Raw Gas						
Production Cost	1.0	1.0	1.0	1.0	1.2	1.2
Value of Liquids	0.4	0.4	0.4	0.4	0.4	0.4
Dry Gas						
Production Cost Net of Liquids	0.6	0.6	0.6	0.6	0.8	0.8
Transmission Cost	0.3	0.3	0.3	0.3	0.3	0.3
Depletion Premium	0.4	0.6	0.7	0.9	1.2	1.5
Economic Price	1.3	1.5	1.6	1.8	2.3	2.6

clearly too large to be ignored, even when reserves are abundant. As discussed in Part IV below, the depletion premium is also important in tariff design, since it is incurred during off-peak as well as peak periods of demand. The approach illustrated in this example is a practical way of deriving an estimate of the depletion premium without heavy additional data or computational requirements.

It also provides a dynamic framework into which the analyst can incorporate different assumptions about future trends in the world prices of substitute fuels, the success of future exploration efforts and the growth of domestic demand. While such assumptions are subject to a considerable margin of error, the framework presented above can be used to test the sensitivity of the derived estimates to alternative assumptions.

3 An Overview of Natural Gas Supply and Demand

For those readers needing a general perspective on the patterns of natural gas production and use across countries, this chapter reviews historical trends and current expectations regarding reserves, production, international trade and consumption of natural gas in industrialized and developing countries. It also reviews the technical terminology associated with the gas operations.

Gas supply and demand are unevenly distributed across countries. As of January 1989, more than 50 per cent of the 116 tcm of proven world gas reserves were in developing countries (including the OPEC countries). However, these countries accounted for only 17 per cent of world gas consumption.[1] The world ratio of proven reserves to current production is about 54 years, as compared with 45 years for oil. There is a great deal of variation across countries from a low of 12.5 years in North America to 154 years in Nigeria, 557 in Iran, and 632 in Qatar.[2] These figures indicate that there is substantial scope for the rapid growth of gas consumption in many countries, especially in the developing world.

The prospects for growth of natural gas production and consumption in developing countries are brighter than those in developed countries. World demand for all forms of energy is expected to grow until the year 2000 at an average annual rate of about 1.6 per cent, which compares with an expected growth rate of about 3.0 per cent per annum in developing countries.[3] In the latter, natural gas is expected to be the

---

[1] *Le Gaz naturel dans le monde en 1988*, Cedigaz, Paris, 1989.

[2] Based on 1988 production and reserve levels.

[3] Mashayekhi, A., 'Natural Gas Supply and Demand in Less Developed Countries', *Annual Review of Energy*, Vol. 13, 1988, Annual Reviews Inc., Palo Alto, California, pp. 119–29.

fastest-growing fuel, with consumption increasing at about 8 per cent per annum. At this rate of growth, as shown later in this chapter, the share of natural gas in total energy consumption in these countries will increase from about 10 per cent in 1988 to about 18 per cent by the year 2000.

3.1 Terminology

Natural gas is a naturally occurring gaseous compound of carbon and hydrogen. Natural gas in the reservoir (raw gas) is a composite product that consists mainly of methane (CH_4). Other constituents of natural gas vary from one reservoir to another. The main constituents of natural gas include ethane (C_2H_6), propane (C_3H_8) and butane (C_4H_{10}), which are referred to as liquefied petroleum gases (LPGs), and pentanes (C_5+), which are heavier hydrocarbons. The gross calorific value of gas increases as it becomes heavier. A natural gas with a high propane content (e.g. greater than 60 bbl/mmcf) is a rich gas. Natural gas can also include other components such as carbon dioxide, hydrogen sulphide, nitrogen and, in some cases, helium and argon. These are undesirable components which must often be removed before the gas can be used. Thus, raw gas is generally separated into its different constituents and purified of inert gases to produce lean gas consisting mainly of methane. This lean gas generally contains less than 20 bbl/mmcf of heavier hydrocarbons. The term natural gas technically refers to all reservoir gas, including methane as well as ethane, LPGs, pentanes and heavier hydrocarbons.

Natural gas exists in the gaseous state or in solution with oil in natural underground reservoirs and is recoverable as gas at typical surface conditions of temperature and pressure. Associated gas is sometimes dissolved in oil and sometimes associated with oil in a gas cap. Associated gas is separated from oil whenever oil is produced at lower pressures and temperatures and in some cases reinjected to maintain reservoir pressure.

The heavier and more valuable parts of raw gas are often liquefied at the surface by well-head separators or gas processing plants. As Table 3.1 indicates, the composition of natural

Table 3.1: Composition of Selected Raw Gases. Percentages of Various Components by Volume.

Component	Groningen (Netherlands)	Leman Bank (France)	Algeria[a]	Ekofisk (Norway)
Methane	81.3	94.8	86.5	85.2
Ethane	2.8	3.1	9.4	8.6
Propane	0.4	0.5	2.6	2.9
Butane	0.1	0.2	1.1	0.9
Pentane	0.1	0.2	0.1	0.2
Carbon Dioxide	0.9	–	–	1.7
Nitrogen	14.4	1.2	0.3	0.5
Total	100.0	100.0	100.0	100.0

Note: (a) These data are for gas dedicated to the LNG project only.

gas differs widely across different fields.

It is important to differentiate between the different categories of reserves. Reserves at any point in time are a fixed stock, but over time this stock changes with new information about economic conditions and with technological development. Countries and companies use different systems of reserve classification, which makes comparison of reserve data difficult. The usual industry practice is to regard *proven reserves* as that fraction of total reserves that geological and engineering data demonstrate with reasonable certainty to be recoverable from known oil and gas reservoirs under present and expected economic conditions and with existing technology. *Probable reserves* are those fractions for which there exists a probability of at least 50 per cent that they will be proven. *Possible reserves* are those quantities of gas thought to be producible with 25 per cent certainty.[4]

In each country the rate of natural gas production is determined not only by the size of its reserves, but also by the size of the domestic market for gas, by the possibility of gas export through pipelines or liquefaction, by the rate of oil production if the gas reserves are associated with oil in the reservoir, and

---

[4] Peebles, M. H., *Evolution of Gas Industry*, Macmillan, London, 1980.

sometimes by other technical or economic factors. The general relationship between gas production and consumption for a given country is:

Production − Flaring − Exports + Imports = Consumption

The difference between the first two terms, production less flaring, is sometimes called marketed production. In the remainder of this chapter we follow this sequence, starting with reserves, to describe global trends in gas production, trade and consumption.

3.2 Natural Gas Reserves

Large natural gas reserves, comparable in size with those of oil, are widely distributed all over the world. The ratio of reserves to production for gas is twice that of oil. Even with expected increases in gas use, its reserves/production ratio is likely to remain above that for oil because the scope for major new gas discoveries is appreciably greater than that for new oil discoveries.

The USSR has the world's largest natural gas reserves by far – nearly 40 per cent of the total. In the Middle East, Iran holds 12 per cent and Abu Dhabi 5 per cent of total world natural gas reserves. Gas reserves by major world region are shown in Table 3.2 and Figure 3.1.

Between 1975 and 1989 total world proven gas reserves increased at an average annual rate of about 5 per cent, from 63 tcm to 116 tcm.[5] Proven gas reserves are expected to increase in the future since there has been a far less systematic search for gas than for oil. In some cases, oil companies that have found gas while searching for oil have abandoned their sites without even declaring their discoveries. Also, when gas is discovered far from a market, new reserves are often not fully delineated. This is particularly true in non-OPEC developing countries where there is a large difference between proven and total potential reserves. In these countries, since the exploratory and appraisal work has remained incomplete, proven supplies are often very small, while diagnostic studies

[5] Cedigaz, 1989.

An Overview of Natural Gas Supply and Demand   33

Table 3.2: World Proven Reserves of Natural Gas. End-1988. Billion Cubic Metres and Percentage Shares.

Region	Volume	% Share
North America	7,994	7.2
Latin America	7,090	6.3
Western Europe	5,512	4.9
Eastern Europe and USSR	42,396	40.0
Africa	7,337	6.6
Middle East	31,235	28.0
Asia	7,694	6.9
Japan and Oceania	2,420	2.2
Total	111,678	100.0

Source: *Le Gaz naturel dans le monde en 1988*, Cedigaz, 1989.

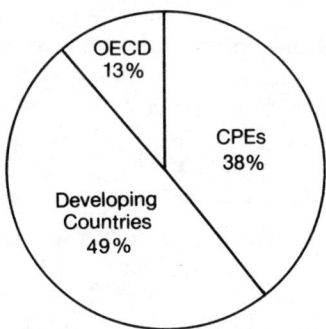

Source: *Le Gaz naturel dans le monde en 1988*, Cedigaz, 1989.

Figure 3.1: Shares of Natural Gas Reserves held by Major World Regions. End-1988.

indicate much larger reserves. This makes the evaluation, costing and planning of gas development much more difficult than it is in industrialized countries with mature gas industries, where supply uncertainty is lower.

Studies by the International Gas Union (IGU) indicate that, because of these factors, potential world supplies of natural gas could increase further, and that natural gas reserves will be sufficient to sustain 90–100 years of production at the projected level of consumption in the year 2000.[6]

Whether gas is found in association with oil or non-associated affects its development prospects. In general, it is easier to develop non-associated gas because of its flexibility, but it is also more costly since associated gas shares some costs with oil.[7] Non-associated gas can be produced gradually or left in the ground until a market develops. For associated gas, unless the market exists or reinjection is possible, the gas must be flared as the oil is produced. About three-quarters of the world's gas discovered so far (but not yet produced) is non-associated.

3.3 Natural Gas Production

Between 1975 and 1988, world gross production and marketed production of natural gas both rose at an average annual rate of 3 per cent to 2.3 tcm. As gas production increased, the ratio of gas production to oil production also increased, and reached 59 per cent in 1988. During the period 1975–88, in developing countries, marketed production grew at 8.0 per cent per annum to 333 bcm, more than twice its 1975 level. During this period the OPEC countries in particular substantially reduced flaring and increased their consumption of natural gas. As shown in Table 3.3, flaring still accounted for a sizeable share of production in 1988, particularly in the OPEC countries.

In 1988, about 209 bcm of total gas production was reinjected. Algeria, the United States, Canada, Venezuela and Iran were responsible for about 80 per cent of total reinjection. In 1988 gas flaring fell to 92 bcm from its 1980 level of 164 bcm, or less than 4 per cent of total production. The OPEC countries were responsible for 49 per cent of the world's total flaring.

---

[6] *Report of the Task Force: World Gas Supply and Demand, 1983–2020*, p. 72, International Gas Union, Paris, 1985.

[7] This is not always the case. For example, Nigeria's associated gas requires an extensive gas gathering system which increases its development cost above that of non-associated Nigerian gas.

Table 3.3: Flaring as a Percentage of Gross Gas Production in Selected Countries. 1973 and 1988.

Country	1973	1988
Nigeria	99	67
UAE	91	8
Iraq	86	42
Saudi Arabia	86	7
Indonesia	84	8
Iran	59	10
Libya	34	21

Source: *Le Gaz naturel dans le monde en 1988*, Cedigaz, 1989.

Developing countries as a group (including the OPEC countries) market only 40 per cent of their total gross production; the rest is flared or reinjected. In developed countries the ratio of marketed production to gross production is far higher at 83 per cent, and in East European countries this share is as high as 97 per cent.

Turning now to the distribution of gas production across countries in 1988, as with reserves, the USSR had the largest production, accounting for nearly 40 per cent of the world's total. It was followed by the United States, with 24 per cent of total production; the Netherlands – a gas exporter – was third with 3 per cent. A breakdown of gas production by region is presented in Table 3.4.

3.4 Natural Gas Trade

Most gas is consumed in the country where it is produced. In 1988, only 13.5 per cent of total gas consumption was based on gas imports, of which over 78 per cent was by pipeline, mainly from the USSR, the Netherlands, Norway and Canada. The rest was imported by sea as liquefied natural gas (LNG). Developing countries – mainly Algeria, Indonesia, Brunei, Libya, and Abu Dhabi – account for almost all exports of LNG (see Figure 3.2). However, the entire LNG trade is only 1.5 per cent of world oil tanker trade.

Between 1975 and 1988, natural gas exports increased at an

Table 3.4: World Natural Gas Production. 1988. Billion Cubic Metres and Percentage Shares.

Region	Volume	% Share
North America	711	30.0
Latin America	127	5.4
Western Europe	207	8.7
Eastern Europe and USSR	859	36.3
Africa	156	6.6
Middle East	153	6.5
Asia	132	5.6
Japan and Oceania	22	0.9
Total	2,367	100.0

Source: *Le Gaz naturel dans le monde en 1988*, Cedigaz, 1989.

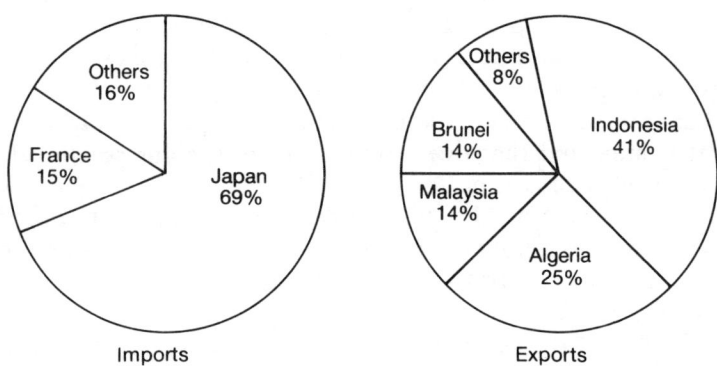

Source: *Le Gaz naturel dans le monde en 1988,* Cedigaz, 1989.

Figure 3.2: Various Countries' Shares of LNG Imports and Exports. 1988.

average annual rate of 6.0 per cent. During 1975–88, total gas imports into Western Europe, the world's largest market for imports by pipeline, rose at an average annual rate of 7.0 per cent. The major importers are West Germany, France, Italy, the United Kingdom and Belgium. About 90 per cent of these imports are through European and Algerian pipelines, with the remainder shipped by LNG tankers, mainly originating from Algeria.[8] Gas imports into Japan, the world's largest LNG importer, grew more than fivefold between 1975 and 1988 and now represent about 70 per cent of total world LNG imports. In the US, gas imports have decreased since 1980. This reflects the stagnation of gas consumption and cancellations of major LNG contracts with Algeria. Canada is the predominant exporter of gas to the US. There has been renewed activity in the US LNG market since 1988 as LNG prices have become more competitive. Norway and Nigeria are both possible new sources of LNG supply to the US in the 1990s.

Developing countries' exports of natural gas more than doubled in volume between 1975 and 1988. The major exporting areas are the Far East and Africa, whose exports rose by four and three times respectively. Indonesia accounts for 41 per cent of total LNG exports, followed by Algeria, whose share is 25 per cent, and, more recently, Malaysia, with a share of 14 per cent. In the Middle East, Abu Dhabi is the only gas exporter, with small exports to Japan. Natural gas trade in Latin America and the Caribbean is limited to the exports from Bolivia to Argentina and from Mexico to the US.

It is expected that over 70 per cent of the gas produced in developing countries until 1995 will be consumed domestically. In view of the low expected growth of energy demand in developed countries and their excess domestic supplies of gas, only a few countries with large low-cost reserves are expected to be potential gas exporters in this century.[9]

The USSR is expected to remain the largest gas exporter. It intends to increase its exports to Western Europe to earn hard currency. In addition to its existing proven reserves, there is much promising offshore territory in other areas. By 1995, Soviet gas exports to Western Europe are expected to double from their 1987 level of 43 bcm, depending on energy demand and decisions about the share of Soviet gas in total gas consumption.

The Middle East's reserves could ultimately prove to be as large as those of the USSR, and provide a major source of gas

---

[8] Libya also has a small LNG contract with Italy.

[9] See Chapter 9 for greater detail.

supply to Western Europe. Iran and Qatar have the largest gas reserves in this region. Iran was becoming a major gas exporter with the IGAT I and IGAT II schemes to export to the USSR.[10] These operations have been shelved, but Iran is going ahead with large reinjection plans in Khuzestan and domestic projects. The offshore North field in Qatar is not yet proven but is expected to be a large field for domestic consumption and potential exports.

In Saudi Arabia and Kuwait, there are large reserves of predominantly associated gas. Further, their large petrochemical and desalination plants and other domestic projects require large amounts of gas. This has led to the consideration of a Gulf gas pipeline network based on Qatar's North field or even LNG imports as well as the development of Saudi Arabia's non-associated gas fields.

In Latin America, Mexico uses a large amount of gas domestically and also exports gas to the US. Other domestic users of gas are Venezuela, Bolivia, Brazil, Colombia, Peru and Argentina. Trinidad and Tobago as well as Chile are contemplating LNG, methanol or fertilizer projects. Meanwhile, given the large and growing regional energy demand, there may be potential for a network across several countries such as Argentina, Bolivia and Brazil.

In Asia, as noted above, Japan is the major importer of natural gas. Japan's consumption of natural gas is relatively small at less than 2 per cent of total world consumption. However, Japan imports over 92 per cent of its gas as LNG, making it the largest importer of LNG in the world. South Korea is importing about 2.6 bcm of LNG per year from Indonesia and is planning to increase its imports in the 1990s. The exporters of gas in this region are Indonesia, Brunei and Malaysia and, since late 1989, Australia. Many other Asian countries, such as Pakistan, Bangladesh, Thailand, Burma, Afghanistan and India, hold substantial gas reserves relative

---

[10] The IGAT I project to the USSR was operating until 1980, and the IGAT II project, which involved gas exports to the USSR and exports by the USSR to Europe requiring net payment by European consumers to Iran, was almost completed but did not come into operation at the time because of the Iranian Revolution. Negotiations are currently under way with a view to resuming the export of gas from Iran to the Soviet Union.

to their domestic energy demand, and use gas in varying amounts in domestic markets.

In Africa, Algeria and Libya are gas exporters. Algeria has a comprehensive plan for the development of gas for export and domestic use. However, pricing and contract disputes led to the cancellation of the El Paso project and suspension of the Trunkline project (both to the US), and several years' delay in the trans-Mediterranean pipeline project to Italy (resolved in 1988). Nigeria has large quantities of both flared and non-associated gas which it is planning to use domestically. The Bonny LNG project will also begin production in the early 1990s. Many countries in Africa have gas reserves that are large relative to their own levels of commercial energy demand but too small for export. With greater urban migration, gas could become an important fuel for industrial uses, power supply and, in some cases, residential and commercial use in Tanzania, the Sudan, Ghana, Somalia, Egypt, Angola, Ethiopia, Tunisia, Morocco, Cameroon and the Ivory Coast. There is also some discussion of a trans-African pipeline for regional gas supply as well as exports to Europe. This possible project would require a pipeline crossing through several countries, with consequent high political risks.

3.5 Natural Gas Consumption

Natural gas accounts for about 20 per cent of the world's total commercial energy consumption. In the West European countries the proportion varies from 15 to 25 per cent. In the US it has remained around 20 per cent for several decades. In several of the developing countries, including Pakistan, Argentina and Mexico, natural gas use is in excess of 25 per cent of commercial energy consumption.

During the 1960s and 1970s there was a rapid growth of gas consumption, largely because the price of gas was competitive relative to oil. As oil and gas prices both rose in the late 1970s, and the growth of total energy demand slowed, the growth in gas consumption also slowed – to an average annual rate of 3 per cent between 1975 and 1988.

The USSR leads natural gas consumption with 35 per cent of world consumption. During the period 1975–85, the USSR

Table 3.5: World Natural Gas Consumption. 1988. Billion Cubic Metres and Percentage Shares.

Region	Volume	% Share
North America	570	29.1
Latin America	81	4.1
Western Europe	259	13.2
Eastern Europe and USSR	784	40.0
Africa	36	1.8
Middle East	93	4.7
Asia	73	3.7
Japan and Oceania	63	3.2
Total	1,959	100.0

Source: *Le Gaz naturel dans le monde en 1988*, Cedigaz, 1989.

increased its consumption at 6.3 per cent per annum. North America was the second-largest natural gas consumer with 24 per cent of world consumption.[11] Western Europe, the third major consuming region with 13 per cent of the world's total, increased its consumption of natural gas at a rate of 3.0 per cent per annum, mainly because of increases in France, Italy and the United Kingdom. In Japan, consumption rose more than fourfold between 1975 and 1988, mainly because of the large volume of LNG imports. Natural gas consumption by major world region in 1988 is presented in Table 3.5 and Figure 3.3.

The largest growth of gas consumption between 1975 and 1988 was in developing countries, where consumption almost doubled to account for 14 per cent of total world gas consumption by the end of the period. Natural gas consumption increased at average annual rates of 16 per cent in Africa, 10 per cent in the Middle East and 5 per cent in Latin America. During the 1990s, developing countries' gas consumption is expected to grow at a higher rate than any other source of

---

[11] US gas consumption rose by about 7 per cent between 1983 and 1984 but fell by 5 per cent in 1985 due to the competition from crude oil prices. It started to increase again in 1987.

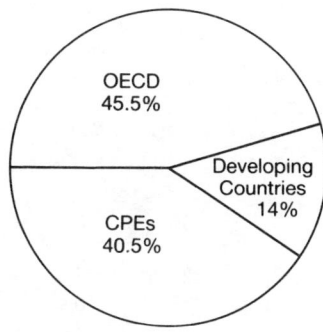

Source: *Le Gaz naturel dans le monde en 1988*, Cedigaz, 1989.

Figure 3.3: Shares of Natural Gas Consumption held by Major World Regions. 1988.

energy at 8 per cent per annum.[12] This is mainly a result of the potential demand from existing energy consumers that could switch to gas.

Because of its qualities as a feedstock and as a clean fuel that can easily substitute for petroleum products, natural gas has penetrated strongly into all energy-consuming sectors except transport. The rate of penetration of natural gas consumption in the various sectors depends on the individual country's economic, industrial and energy situation as well as its demographic and climatic conditions. Table 3.6 shows the percentage breakdown of natural gas consumption by sector for 1986.

The largest demand is in the industrial market, which is generally the first stage of diversification in the development process of the gas industry. In 1986, according to Cedigaz, this market accounted for 28 per cent of world gas demand, and its share is expected to increase along with the use of gas as a raw material for the production of fertilizers and methanol, which currently account for only 6 per cent of total consumption.

The power market accounts for 25 per cent of total gas consumption and this share is expected to grow with the use of

---

[12] Mashayekhi, A., ibid.

Table 3.6: Structure of World Natural Gas Consumption by Sector. 1986. Percentage Shares.

Region	Power Plants	Industry[a]	Raw Material	Residential and Commercial[b]	Other[a]
North America	4	29	4	40	13
Latin America	18	36	6	10	30
Western Europe	13	27	6	48	6
USSR	39	28	7	14	12
Eastern Europe	25	34	10	17	14
Africa[c]	33	13	3	4	47
Middle East[c]	29	25	9	6	31
Japan	73	2	1	–	24
Australasia	32	35	6	14	13
South East Asia	19	32	2	8	20
World	25	28	6	26	15

Notes: (a) Excluding raw material.
(b) Includes gas used in the transportation and agricultural sectors.
(c) Includes gas used in production, treatment and processing.
Source: *Le Gaz naturel dans le monde en 1988*, Cedigaz, 1989.

new gas turbine technology. Japan is an exceptional case where 75 per cent of gas use is in the power sector, mainly for environmental reasons. In other developed countries, the share of gas in power is lower, for example, 14 per cent in the USA and 12.5 per cent in Western Europe. This sector is large and growing in many developing countries with less diversified economies; it often accounts for over 30 per cent of gas demand.

Residential and commercial demand for cooling, water heating and space heating is also large, particularly in developed countries, and accounts for 26 per cent of world gas consumption. Climate, housing density and the level of income are the main determinants of gas penetration into the residential and commercial sectors. In many developed countries, the residential gas market remains the major growing market. The commercial market largely covers gas use in offices, hotels, schools and shops. In developing countries with warm climates, residential and commercial gas use will be relatively small. However, in countries with space heating, such as Pakistan, the residential and commercial sector accounts for about 15 per cent of gas demand.[13]

The figures in Table 3.6 hide the large differences in the importance of gas use to the various sectors. In 1986 in the US, gas provided over 40 per cent of energy used in the residential and commercial market. In the industrial sector, gas accounted for 29 per cent of total energy used. In the power sector, the share of gas was 14 per cent.

The sectoral demand pattern in Western Europe is similar to that in the US. In Western Europe in 1986, gas accounted for about 37 per cent of energy consumed in industry, 12.5 per cent in power generation, and 48.5 per cent in the residential and commercial market. As gas supplies have become more expensive, the share of gas in the residential and commercial sector – where the value of gas is higher – has increased.

In the CPEs the pattern of gas consumption is heavily skewed towards the power and industrial markets. Gas provides about 14 per cent of residential and commercial energy in the USSR, 28 per cent of industrial demand and 39 per cent of power generation.

The pattern of gas use is very different in developing countries, where a large proportion of gas (up to 90 per cent) is used in the industrial and power sectors. In Pakistan, for example, about 35 per cent of gas consumption is in the power sector, 35 per cent in industry, 16 per cent in fertilizer and 14 per cent in the residential and commercial sector.

3.6 Conclusion

The primary objective of this chapter has been to provide an overview of the patterns of natural gas reserves, production, trade and consumption across countries. However, it is also possible to draw some general conclusions from this review about the future role of natural gas in meeting the world's energy needs.

It is clear from the figures presented above that the world's

---

[13] For additional detail see Chapter 9 below.

current proven natural gas reserves are more than sufficient to meet the anticipated demand for gas over the remainder of this century and for several decades beyond. Moreover, it is quite likely that large additional reserves could be found if there were sufficient demand. There is scope for increased domestic consumption of gas in the developing countries, and more than enough potential candidates among those countries for LNG export. While the world may be running out of oil, this is not the case with natural gas.

The issue, therefore, is not the physical or technical constraints relating to reserve size and production characteristics, but rather the economic considerations deriving from the location of reserves relative to potential markets, the costs of producing and transporting gas relative to the costs of alternative fuels, the value of gas to potential users in the different market segments, the price structure facing potential consumers and producers, and the general economic and political environment in both the producing and the consuming countries. In the rest of the book we examine these economic parameters from both theoretical and practical perspectives.

PART II

THE COSTS OF GAS DEVELOPMENT

4 The Economic Cost of Gas Development

A number of theoretical and practical questions are involved in calculating the cost of gas for a field or for a country. This chapter decribes how gas costs are derived and discusses some of the complexities that may be encountered. Chapter 5 provides estimates of the marginal costs for eight developing countries. A detailed case-study of the estimation of marginal costs and related tariffs for Bangladesh is provided in Chapter 11.

The marginal cost of natural gas supply is a necessary input to a number of important investment and pricing decisions. First, the cost of gas relative to other fuels is needed to decide questions of fuel choice. Second, benefits net of costs can be compared in different uses in order to determine an optimal allocation or utilization pattern for gas. Third, the relative costs of producing and transporting gas from different fields need to be known if a field development strategy is to be planned effectively. Finally, gas supply costs are an important component of prices and a necessary input for negotiating prices with producers, transmission companies and consumers.

4.1 Components of Gas Systems

A gas supply system can be divided into four interrelated parts: exploration, development and production, transmission and distribution. Unlike oil, gas can rarely be used immediately after production. Consumption of the first increment of gas generally requires a large initial expenditure on a whole network of exploration, production, transmission and distribution facilities. Thereafter, the consumption of additional volumes necessitates little additional expenditure until the system's maximum capacity is reached. Because of this lumpy investment pattern, one of the greatest sources of difficulty in the analysis of gas supply costs is that these steps are all interrelated, and a whole gas infrastructure up to the burner-tip should be planned in an integrated way. These problems do not apply in the case of oil.

There is a large drain on cash flow in the initial period of expenditure on exploration, development, transmission and distribution. The whole system covering these four phases can take anywhere from five to ten years depending on the technical complexity of the field, the size of the system, the economics of the project and the management capabilities of the institutions in charge of its implementation.

Exploration activities aim at discovering whether oil or gas fields are present in a particular area and, if present, whether they are sufficiently large and productive to be worth developing. Usually, exploration begins with a study of existing geological information followed by geological and geophysical surveys. These determine whether the drilling of exploratory wells is justified. A typical onshore well will cost $10–15 million while the cost of an offshore well is likely to be in the range of $15–30 million.

Apart from the purely technical costs of exploration, one must take account of the risk. There are many ways of doing this.[1] The basic idea is that, for any given company, the costs and probability distribution of finding reserves are averaged out over a number of possible basins. In most exploration models, there is a relationship between effort and yield. In some models the predicted yield per unit of effort is assumed to remain at historical levels. This can lead to a serious underestimation of costs if there are diminishing returns to effort. Generally, when a new field is found, the yield is high since the bigger reserves are found first. Yields usually taper off over time as an area is more intensively explored. This decline is depicted as a non-linear process in some models.

One complication in estimating the exploration costs of gas development is that gas is often found by oil companies that are drilling for oil. This makes the division of costs between oil

---

[1] For a survey of various modelling approaches to projecting the volume of oil and gas to be discovered see Adelman, M. A., Houghton, J. C., Kaufman, G. and Zimmerman, M. B., *Energy Resources in an Uncertain Future*, Ballinger, Cambridge, Massachusetts, 1983.

and gas difficult.[2] When gas is found in the search for oil, all the exploration costs can be attributed to oil. There are, of course, gas-prone zones where companies search in the knowledge that they will find gas.

The second part of a gas supply system is *development and production*. This covers development drilling, field preparation, field gathering, compression, separation of natural gas liquids and treatment of the gas. Its aim is to produce pipeline-quality gas to meet contract volume, quality and pressure requirements. Development and production costs are essentially a function of the number of wells that are necessary to develop a field, their location (their distance from the market, and whether they are on or off shore), the condition of the reservoir, and the surface infrastructure required (for gathering, separation, treatment and metering). The latter depends largely on the distance of the wells from one another and the quality of gas. Total development and production costs are an increasing function of the number of wells and the volume of production, but a declining function of productivity and the quality of the gas.

The allocation of development and production costs between oil and associated gas and, similarly, between lean gas and other gas by-products such as LPGs and gasolines is difficult. The recovery of associated gas often depends on oil production, and the need for technical efficiency in oil recovery can dictate the amount of gas produced. In some cases, it is necessary to reinject the gas to increase oil recovery, thereby delaying use. The appropriate cost allocation depends on the particular circumstances of each case.

Natural gas and crude oil are often discovered and produced by the same company using the same inputs. Because of this, it is reasonable to think of oil and gas as joint products. The basic theoretical treatment of this problem resembles that for a multiple-product firm. The marginal cost of each output should be estimated separately by keeping the quantity of other outputs constant while the production of one output is increased by one unit. In practice, a major problem is the lack

---

[2] See the discussion later in this chapter on joint costs.

of sufficiently detailed cost data. A number of relatively complex models have been developed to deal with the joint cost problem. For example, in the case of US gas in the 1960s, the Federal regulators' decision was to include an amount in the ceiling gas price that would compensate gas producers 'sufficiently' for their efforts without specifying what a 'fair' profit was. None the less, there is no universally accepted solution to the problem of joint cost allocation between oil and gas.[3]

There are three relatively simple – although not entirely satisfactory – methods of allocating joint costs. These are the allocation of costs: (a) according to market value; (b) according to calorific value; and (c) by subtracting the benefits of by-products from total costs and allocating the remaining costs to the main product. Since, in many cases, the market value of gas is not known in advance, the first method often can not be used. The problem with the second method is that the calorific value of joint products is not necessarily related to their market value and thus this allocation becomes arbitrary. The third method is most often used since the market value of by-products (e.g. gasoline, LPGs, etc.) is easily observed and thus can be subtracted from total costs to derive the net costs of lean gas production. This method is particularly suitable in cases where the gas is the main product and the other products are by-products rather than joint products.

The third stage of gas development is the *transmission* of gas from the field or treatment plant to the city gate. Investments in transmission facilities are lumpy and costs are subject to significant economies of scale. These costs depend primarily on the diameter of the pipeline, which is a function of peak demand, as well as the length of the pipeline. An average incremental cost method (as described below) is generally used for their estimation.

The fourth stage is the *distribution* of gas to end-users. Distribution costs may differ depending on the size and pattern of customer demand. While the cost of delivering gas to a major industrial power plant is large, the incremental cost per cubic

---

[3] For a review see Walters, A. A., 'Production and Cost Functions: An Econometric Survey', *Econometrica*, 1963, pp. 1–66; Griffin, J. M., 'Joint Production Technology: The Case of Petrochemicals', *Econometrica*, Vol. 46, No. 2, March 1978, pp. 379–97.

metre is quite low because of the economies of scale involved. On the other hand, the cost of a distribution network for the residential and commercial sector is quite high because the size of demand by this sector is relatively small. Investment in distribution must be based on the peak demand in the system. In general, distribution costs are sensitive to the average and peak volumes of demand, the density of population, and the technical characteristics of the system (e.g. whether it has plastic or metal pipelines). Again, average incremental cost methods are generally appropriate.

4.2 Cost Estimation

The four components of an integrated gas system, as described above, have different cost characteristics. The most important of these stem from their differing economies of scale and lengths of construction period. In Chapter 10 we also discuss how different cost components should be allocated to different types of consumers (e.g. peak and off-peak).

For financing decisions the total project cost is important, including provision for inflation and working capital during the construction period. For economic decisions, however, costs are considered on a cash-flow basis (i.e. when they are incurred, without regard as to how they are financed) and expressed in 'real' currency units of some base year to exclude inflation. Where costs are incurred over several years, a discount rate can be used to express each year's cost in terms of its 'present value' in the base year. In this way the 'present value cost stream' can be summed over the construction period to yield a total economic cost of the project.

For most types of economic analysis, the unit cost – rather than the total cost – is needed. There are two types of unit cost: average and marginal. The average cost is simply the total cost of the system, or of a component, divided by the total gas produced (or transmitted, distributed, etc. through that component). For example, if the total economic cost of a new gas development project were $500 million and it could supply 12.5 billion cubic feet (bcf) per year for a twenty-year period, then the average project cost would be $500 million/ (12.5 bcf × 20) = $2/mcf.

The average cost approach is not appropriate for any cost component that has significant economies of scale. This is because the costs and benefits of such a component will occur over different time-paths. Most costs will be incurred in the early years (when there may still be no production), while the benefits will materialize only over time (when there may be few costs). In general, the only gas system component for which average unit costs will be appropriate is exploration. Because of the risk element, these costs cannot be precisely defined in advance in any case, so it is appropriate to use an estimate based on historical experience with unit costs and success ratios. Examples of this approach are presented in the next chapter.

For production, distribution and transmission cost components an estimate of marginal, or incremental, costs is needed. As described in Chapter 10, marginal costs are also the appropriate basis for pricing. Where economies of scale are not important, average costs and marginal costs will be similar,[4] but where there are scale economies, it becomes particularly important to focus on marginal, rather than average, costs.

There is a major debate in the economic literature over whether long-run or short-run marginal costs should be the basis for pricing when there are economies of scale in production. On theoretical grounds, short-run costs are to be preferred, but in the real world of consumers with limited information and uncertainty about the future, long-run marginal costs are generally the more practical choice. In the case of natural gas, where economies of scale are a major determinant of investment size and timing, long-run marginal costs are the most useful parameter for planning and other types of economic analysis.

The most satisfactory and widely-used method for calculating long-run marginal costs is the average incremental cost (AIC) approach.[5] The AIC is estimated by dividing the discounted incremental costs of meeting future demand by the

---

[4] In the hypothetical case of constant returns to scale, they will be identical.

[5] For a review see Munasinghe, M., Warford, J. J. and Saunders, R., 'Marginal Cost Pricing', Staff Working Paper, World Bank, 1974.

discounted volume of incremental output over the same period. The numerator in the *AIC* formula is the present value of the least-cost investment stream plus the incremental operating and maintenance cost. The time-stream of expenditure for providing, maintaining and running the system (in the numerator) must correspond to the production stream over time (in the denominator).

The *AIC* formula is shown below:

$$AIC = \frac{\sum_{t=1}^{T}[\{I_t + (R_t - R_0)\}/(1 + r)^t]}{\sum_{t=1}^{T}[(Q_t - Q_0)/(1 + r)^t]} \quad (4.1)$$

where I is the marginal capital cost, $R_t - R_0$ is the marginal operating and maintenance cost due to the new demand, $Q_t - Q_0$ is the marginal demand, and r is the discount rate. In the next chapter we show how the *AIC* approach can be used to compare gas costs in widely different circumstances.

5 Gas Development Costs in Eight Countries

The last chapter described the principles behind marginal cost estimation. This chapter presents the results of applying that framework to eight developing countries.[1] These countries represent a range of gas sector situations: large and small reserves, offshore and onshore gas, associated and non-associated fields, gas with different compositions, diverse field specifications (shallow and deep structures), and long and short distances to the market.

The quantity and quality of data vary for the different countries. In some cases a detailed and reliable data base exists for every field. In others results are based on information on one or two fields. Most of the data are from World Bank-financed projects or consultant studies.

Generally, costs are estimated on a field-by-field basis for most major fields in each country. This basic approach is designed to enable us to analyse the relationship between costs and output for an entire producing region. Thus we treat output expansion as occurring by discrete increments corresponding to the start of production at individual fields. Field development ideally proceeds from the lowest-cost gas to more expensive gas, taking into account the differences in the costs of exploration, production and transmission to the market. In practice, because of particular contractual agreements or incomplete knowledge, this process may not take place. The discovery of new lower-cost fields over time may also change the original ordering of fields.

5.1 Incremental Costs of Gas Development

The present values of the costs of gas production are shown in

---

[1] This exercise is carried out in greater detail for Bangladesh in Chapter 11.

Table 5.1: Present Value Exploration, Development and Transmission Costs and Projected Volumes of Natural Gas Production in Eight Countries. 1986–2005. Costs in Million 1982 Dollars. Volumes in Billion Cubic Feet.

Country	Cost	Volume
Cameroon	514.9	287.7
Egypt	1,847.4	2,641.5
India	2,001.8	1,870.1
Morocco	360.1	210.4
Nigeria[a]	1,889.5	1,717.7
Tanzania	184.7	175.9
Thailand	3,716.3	2,396.8
Tunisia	773.4	483.3

Note: (a) Excludes exploration costs.

Source: Authors' calculations based on World Bank data.

Table 5.1. The first column shows the full incremental cost of that gas delivered 'at the city gate' over the period 1986–2005. It includes all exploration, production and transmission costs, as well as the costs of compression facilities that maintain pressure where necessary. These costs are close approximations to the costs of gas delivered to major industrial, power and fertilizer users that are located on or close to the transmission grid. Distribution costs to serve smaller commercial or residential users are highly site-specific and have not been estimated.[2]

In this sample, exploration costs are small relative to development and transmission costs. This is because in these countries gas is frequently found during the search for oil and there is no clear indication of actual expenditure allocated to gas. None the less, an estimate of future exploration costs is included in the cases of all countries except Nigeria, whose large proven reserves were already sufficient (in 1986) to last until beyond the year 2005.

In some countries, such as Tunisia and Morocco, it was possible to estimate the costs of future reserve additions on the basis of a planned exploration effort. In others, historical cost

---

[2] See Chapter 12 for an example of the estimation of distribution costs.

data, together with success ratios, the estimated probability of successful plays in the future and the expected average discovery size, were used to determine future exploration costs.

The current and future development costs for known and proven fields include field preparation, drilling, and separation and treatment facilities, to provide a pipeline-quality gas. The cost of future drilling and field compression to maintain the pressure of gas production over the twenty-year period are also included.

The transmission costs cover offshore and onshore pipelines to move the pipeline-quality gas to the city gate of the major markets. These are based on country data or estimates made by engineers of the costs of pipelines, telecommunications, surveying and rights of way. In some cases, such as Thailand and Nigeria, they include the compression facilities needed to maintain pressure over the twenty-year planning period.

5.2 Expected Incremental Demand and Production Profiles

The estimates of discounted production levels, shown in Table 5.1 as the present values of projected volumes of gas production, go beyond a single-project approach and are based on the long-run demand. These projections are adopted as the production profiles except where production is limited by the supply constraint. In the latter case, the production profile will depend on the potential physical supply of gas. About half the countries in the group are demand-constrained or gas-surplus countries; there is a positive gap between their available long-run natural gas supply and domestic demand. The other half are supply-constrained or gas-deficit countries. The forecasts of their long-term demand exceed their long-run supply potential over the period 1986–2005.

These production estimates are based on potential demand scenarios for natural gas. Their realization depends on: (a) the relative prices of other energy sources; (b) the degree of government commitment to natural gas development; (c) the institutional strength of the organization in charge of gas development, transport and marketing; and (d) the availability of appropriate financing. Because of economies of scale, higher demand estimates would reduce the marginal cost and lower estimates would increase it.

These demand estimates are based on a static analysis. In a dynamic system, a large gas discovery could affect national income and the energy elasticity and bring about a shift in the whole demand curve. Generally, once gas is discovered in large quantities, plans to use it in new plants also emerge. Therefore, the estimates used in the study provide a lower bound to potential future demand. Increased demand will lead to increased production in countries with a demand constraint and reduce the marginal cost of gas supply until capacity is reached.

5.3 Results

Table 5.2 shows the average incremental costs of gas at the well-head and at the city gate based on the figures in Table 5.1. In all cases, the long-run marginal cost of gas delivered at the city gate is far below the costs of alternative fuels or feedstocks. The production costs range from $0.61 to $1.29 per

Table 5.2: Marginal Costs of Natural Gas at Well-head and City Gate in Eight Countries.[a] Constant 1982 Dollars.

Country	Well-head		City Gate	
	$/mcf	$/boe	$/mcf	$/boe
Cameroon[b]	1.29	7.60	1.79	10.54
Egypt	0.65	3.81	0.71	4.18
India	0.95	5.60	1.51	8.88
Morocco	1.16	6.48	1.71	10.07
Nigeria[c]	0.65	3.83	1.10	6.48
Tanzania	0.61	3.59	1.05	6.18
Thailand	0.80	4.71	1.50	8.84
Tunisia	0.67	3.97	1.60	9.43

Notes: (a) These calculations assume a discount rate of 10 per cent, and exclude all profit, tax, royalty and depletion costs.
(b) Production for the domestic market; inclusion of exports would reduce costs.
(c) Includes both associated gas at $0.82/mcf and non-associated gas at $0.44/mcf (at the well-head)

Gas Development Costs in Eight Countries 59

mcf, or from $3.59 to $7.60 per barrel of oil equivalent (boe). Moreover, the marginal cost of gas is not expected to rise in most of these countries in the period 1986–2005 as they tap their proven stocks of reserves. However, as the reserves/production ratios become smaller, the marginal costs are expected to rise as new, more expensive fields are discovered and developed.

The most important determinants of the costs of gas production are: (a) the size of reserves; (b) the rate of production per well and the rate at which production begins to decline; (c) the location of fields relative to centres of demand; (d) the composition and pressure of the gas; and (e) the level of demand. Since in each of our sample countries there is a proven stock of gas, the exploration costs (estimated on the basis of the probability of success) are a relatively small proportion of total costs.

The production costs of large onshore fields lie at the lower end of the cost spectrum. Their relatively low drilling costs often result partly from the concentration and productivity of wells. The costs of production from onshore fields increase with the depth of wells.[3] In Morocco some wells are over 13,000 feet deep. This, together with low recoverability, increases production costs. Another factor that increases costs is the dispersion of gas fields. Gathering costs depend on the number of wells and the distances between them. In Cameroon, the costs of an extensive gas gathering system from a large number of wells over a wide area increase the costs of gas supply.

Production and transmission costs for offshore fields are generally far above those for comparable onshore fields. The offshore location of some fields in India, Thailand and Tanzania leads to higher drilling and operating costs and higher *AICs* than they would have with only onshore fields. This is due to both the higher cost of equipment and the greater difficulties involved in offshore operations.

The size of reserves and productivity per well affect costs. Thus in Tunisia, for example, the larger fields such as Miskar

---

[3] Deeper wells have to be wider at the top and require special equipment to deal with the higher pressure encountered.

benefit from economies of scale in production in comparison with other, smaller, offshore fields. Transmission costs are also higher off shore because the costs of the pipeline are greater. These costs depend on both the volume of throughput and the distance of the field from the shore.

With a given size of reserves, the cost profile generally falls and then rises over time. At the beginning, the production/reserve ratio is low, and the marginal cost of production is falling as production rises. Once the plateau production level has been reached, compression is generally required to increase the pressure of the delivery system. Compression provides relatively small additions to throughput but increases the cost of the incremental production. Then, as demand increases further, more expensive fields have to be brought into operation. Obviously, if a large new discovery is made it can reduce costs.

Tanzania is an example of a country that is only beginning to consider gas development and use. Costs are falling since exploration and development wells already drilled at Songo-Songo are sufficient to meet an increase in demand up to 100 mmcf/d for twenty years with some compression after 1990. Further exploration in other areas such as Mnazi Bay, which is further from the market, is expected to increase the potential production of gas.

Indivisibilities and increasing returns, with a high reserves/production ratio, followed by diminishing returns in providing gas at a uniform rate as the ratio of reserves to production declines, give rise to a U-shaped marginal cost curve. Most developing countries are still in the initial stages of gas development and therefore in the falling portion of the curve.

Nigeria is the only country considered here whose gas development is based on reserves of associated gas. In general, costs are presumed to be lower for associated gas than for non-associated gas, because a large share of its exploration and development costs are joint costs and are partly allocated to oil. In the case of Nigeria, however, although associated gas is collected at no cost at the flare point, it needs to be gathered from low-volume, widely-dispersed supply points, and then treated and compressed. The cost of gathering, treating and compressing Nigeria's associated gas is $0.82/mcf, compared with only $0.44/mcf to produce and treat the gas from Nigeria's large non-associated fields.

Depending on the composition of the gas, purification, processing or separation may be necessary. The richness of the gas is also important as well as the impurities such as hydrogen sulphide, nitrogen and inerts. In some gas fields in Tunisia, for example, the gas is of a low quality, and there are additional costs in processing it before it can be used.

The estimates given in Table 5.2 have not been adjusted to take into account the composite nature of the natural gas, which includes dry gas as well as NGLs in different proportions. Thus, these costs may overestimate the cost of lean gas. If joint costs were allocated partly to oil, the cost of producing dry gas would fall. Joint cost issues are more important in Thailand and Tunisia, where gas has a high NGLs content, than in Tanzania where the raw gas is already fairly dry. For a field in Thailand, if costs are allocated proportionately according to the heat content of each product, the dry gas costs are about $0.20/mcf below the unadjusted costs given in Table 5.2. Similarly, the costs of lean gas in Tunisia are estimated to be about $0.15/mcf below the unadjusted costs.

In many developing countries with large gas reserves, gas production is constrained by the size of the market. The actual production profile is therefore far below that warranted by the technical potential of the reserves. Hence, the overall gas system does not benefit from economies of scale as much as, for example, the system in Pakistan, which operates close to production capacity. The sensitivity of the costs to the size of the market is a major issue in several small African countries where potential markets are limited. In gas-surplus countries, as the market expands, the benefit of economies of scale in production and transmission becomes apparent.

Costs are very sensitive to the length of each field's productive life. In some cases it is possible to increase annual production rates considerably by shortening the field's productive life. Because of the economies of scale in gas development and transport, the increase in costs may be less than proportionate to the increase in production. Although twenty years for field depletion is often used by engineers as a rule of thumb for planning, it may be sensible to use up the gas over a shorter period when there are users who can easily switch to other fuels later (see Chapter 2 for a discussion of depletion costs).

5.4 Conclusions

The data from these eight countries indicate that the marginal cost of gas is especially low in cases where large indivisible investments have been made and total production is a small proportion of the volume of gas reserves in place. Costs may rise when production reaches levels high enough to require additional compression facilities, a new parallel pipeline, or increased expenditure on exploration to replace the gas that has been produced. Most developing countries are currently in the falling part of U-shaped marginal cost curve.

The level of marginal costs differs from field to field. Deep, low-pressure reserves have higher costs than shallow, high-pressure fields. Production from onshore fields for a larger market results in lower costs than offshore production for a small market. Despite variations across countries, it is clear that the economic cost of natural gas delivered at the city gate lies far below the cost of alternative fuels. Costs are expected to remain low for the next 15–20 years in many developing countries, while their reserves/production ratios will remain high. This is an encouraging result for countries embarking on the development of their gas reserves. Moreover, as systematic exploration proceeds, reserves and production estimates are likely to be revised upwards.

PART III

THE BENEFITS FROM GAS DEVELOPMENT

6 Gas Benefit Valuation

In Chapters 4 and 5 we considered the theory and empirical evidence on the cost of gas development. We now deal with the benefit side. In this chapter we discuss the techniques of benefit valuation for natural gas uses, the meaning of the netback concept and the circumstances under which this concept will and will not provide an appropriate measure of benefits. Chapters 7–9 describe the application of benefit measurement to three distinct markets for gas: LNG export (Chapter 7), power and industry (Chapter 8), and residential and commercial distribution (Chapter 9).

Much of the empirical work described in these chapters originated in a set of consultant studies commissioned by the World Bank and reported more fully in working papers, references to which are given in the relevant chapters. The calculations reported here were revised in 1988–9 for the purposes of the present study. In commissioning these studies, standard assumptions were given to all the consultants and the methodologies were defined to yield a value of gas *at the point of delivery to the user*. The basic idea was to begin with the value of the end product to the consumer (e.g. fertilizer, electricity, 'burner-tip' gas for cooking and heating, or regasified LNG at the point of entry into the importing country's grid) and to net out all the costs involved in producing and delivering that end product. Thus the gas benefits in this part are calculated in a symmetric way to the gas costs of the previous part, where those costs included exploration, production and transmission of gas *up to the point of delivery* to a large project or distribution system.

The calculations in all the chapters in this part are based on constant 1982 dollars and mid-1989 fuel prices escalated at 2 per cent per annum in real terms. The base-case discount rate is 10 per cent in real terms and inflation is excluded from all costs and benefits. Sensitivity tests are used to measure the effects of different assumptions. The purpose of these empirical models is to illustrate the techniques of benefit measurement and the factors on which the results hinge, rather than to provide numerical answers that can be applied directly to other cases.

The use of gas for industry and power is almost universal in developing countries with gas resources, but the value of gas in these markets varies widely across countries. Site-specific factors, such as whether the country is large enough to absorb the output of a world-scale ammonia/urea plant, or whether the alternative to gas-fired power generation is hydropower or small oil-fired turbines, are clearly of major importance. For these reasons we spend little time in Chapter 8 on the cross-country pattern of gas use in those sectors or on individual empirical estimates. Rather we describe the types of possible benefit measurement that will be appropriate and present the results of a simulation exercise that shows how various factors affect the value of gas in the power sector.

The use of gas for residential and commercial distribution and for export after liquefaction are much less common and are clearly not of universal relevance for the developing countries. Therefore, in Chapters 7 and 9 we begin by reviewing the existing state of the markets for LNG export and residential and commercial distribution before proceeding to the empirical model for measuring benefits.

6.1 Why Measure Gas Benefits?

Calculations of the benefit derived from gas use are needed for three main reasons. First, at the project level, the benefits from a particular gas use are sometimes compared with the costs of gas development as a measure of a project's profitability. Second, in a gas-short country, estimates of the benefits from competing uses of gas are needed in order to allocate supplies to their highest-value use. Finally, in evaluating the options for a gas pricing policy, the benefits from different uses set an upper limit on the price that can be charged, based on the consumer's willingness to pay.

While gas benefit estimates are an essential part of these three types of analysis, they do not provide the full answer. At

Gas Benefit Valuation

the project level, for example, finding that gas benefits exceed costs for a potential project does not necessarily mean that the project should be endorsed. There may be alternative projects whose benefits would be even higher. Similarly, the comparison of benefits from two competing gas-using projects cannot, by itself, determine which should be preferred unless the two are identical in scale and timing and have no links to other potential gas projects – a situation which rarely obtains in a small gas market.[1] For gas pricing, as discussed in Chapter 10, in addition to practical problems of implementation, it will rarely be in the country's best interest to price as a discriminating monopolist who tries to capture the entire consumer surplus by setting the price for each user at his maximum tolerance.

6.2 Defining the Benefits Attributable to Gas

Benefit valuation is a parallel, but entirely independent, exercise from the gas costing discussed in Part II above. There we were tracing out the supply curve for gas; here we are mapping the demand curve. The benefits we are measuring represent the full area under the demand curve, taking no account of the cost of supplying the gas to the user. Thus the presence of positive benefits does not necessarily indicate a project's viability. The *net* benefits of gas can only be quantified once both costs and benefits have been estimated, as described in Chapter 2.

Benefits attributable to gas are defined as the excess of overall project benefits when gas is used in preference to the next-best alternative. This general definition is best illustrated by referring to the three possible cases: (a) when the gas replaces another fuel or feedstock in domestic production; (b) when the gas is used to produce something that would otherwise be imported; and (c) when the gas is directly or indirectly exported.

---

[1] Chapter 12 describes the appropriate framework for using gas benefits along with other information to compare competing projects.

(a) Substitute Fuel or Feedstock. The most straightforward case is when gas is used in place of an alternative fuel or feedstock for domestic production of, for example, cement or fertilizer. In this case, the total benefit attributable to gas B_g is equal to the total cost of the alternative fuel or feedstock F_a over the life of the project less the difference in investment or conversion costs of the two processes, $I_g - I_a$:

$$B_g = \sum_{t=1}^{T} [\{F_{at} - (I_{gt} - I_{at})\}/(1 + r)^t] \qquad (6.1)$$

It is often more convenient to express the gas benefit in terms of a benefit per unit of gas consumed over time. This is sometimes referred to as the 'netback' value of gas in that use.[2] The netback N is calculated by dividing B_g (in constant price monetary units) by the present value of the gas consumed Q_g, over the life of the project (in thermal or volumetric units):

$$N = \frac{\sum_{t=1}^{T}[\{F_{at} - (I_{gt} - I_{at})\}/(1 + r)^t]}{\sum_{t=1}^{T}[Q_{gt}/(1 + r)^t]} \qquad (6.2)$$

As defined above, the gas netback could also be called the average incremental benefit (*AIB*) of gas use. It is closely analagous to the definition of average incremental cost set out in Chapter 4. Both are long-run concepts, using the full benefit or cost streams over the life of the gas-using or gas-producing project. Both are based on *actual* gas consumption or production as they develop over time, not the capacity of the project. In a simple case, if a single gas-producing project were built to serve a single gas-consuming plant, then the denominators of the *AIB* and *AIC* would be identical. In this case the *net* benefit stream $(B_g - C_g)$ would be the net present value of the project pair. Even in the more general case, a rough initial idea of a project's likely viability can be established by comparing its netback value with the *AIC* of gas in

---

[2] This use of the term 'netback' is quite distinct from the concept of netback pricing of crude oil. The only similarity is that both derive a value of an input (crude oil, gas) from the value or price of the end product.

that country or region. If the *AIC* is higher than the netback, the project is unlikely to be economically viable.

While the netback value of gas in a certain use is a handy summary measure, it also has some limitations. These arise mostly because it is a measure of the *average*, not the *marginal*, value of gas in that use. Thus the netback is not generally a good guide for pricing. In theory the netback represents the *average* price which, if maintained over the life of the project, would just cause the project to break even compared to the next-best alternative. However, in practice, if the gas price were set equal to the netback, those consumers with downward-sloping demand curves would switch out of gas before the change was economically justified.[3]

Another problem with the netback arises when gas is a relatively small input into the total cost of a project. For projects that are highly capital intensive, such as petrochemical or fertilizer plants, the netback value of gas is extremely sensitive to operating assumptions such as capacity utilization rates. Because such factors affect overall project profitability more significantly than does the cost of gas, the netback value of gas in such a use can be quite volatile. It has been estimated that for a fertilizer plant, for example, the effect of operating at 70 per cent rather than at 90 per cent capacity utilization is equivalent to having to pay an additional $2.00/mmbtu for the gas used by the plant. Although the netback calculation is still a valid one in such cases, the estimate that results should be treated with caution.

(*b*) *Import Replacement*. Sometimes the availability of gas makes possible the production of goods that would otherwise have to be imported. For example, a small country without gas resources would probably find it more economic to import its fertilizer requirements than to set up local production based on naphtha or fuel oil. If natural gas were discovered in such a country, the value of gas for fertilizer production should be calculated from the import savings it would bring, not from the hypothetical replacement of gas by naphtha.

---

[3] This problem would not arise for very small increments of gas or over any 'flat' sections of the consumers' demand curves, since there the marginal and average benefits would be equal.

In the import replacement case, the total benefit from gas is represented by the import value of the production M, less the capital and non-gas operating costs of the plant over its lifetime:

$$B_g = \sum_{t=1}^{T} [(M_t - I_t - O_t)/(1 + r)^t] \quad (6.3)$$

The netback value would be calculated as before, by dividing B_g by the present value of the gas consumed over the life of the project. This would provide the average incremental benefit per unit of gas consumed.

(c) Direct or Indirect Export. Gas is indirectly exported when it is used to produce a commodity for the export market. Several of the OPEC countries have built large ammonia and petrochemical plants, using their associated gas as a feedstock, in order to supply the world market. In this case the total benefit from gas use is calculated exactly as in the import case, except that the value of the commodity (M in equation (6.3) above) is its export (f.o.b.) price instead of its import (c.i.f.) price.

Sometimes gas-using plant will be built in developing countries whose output is partly consumed domestically and partly exported. It is a straightforward matter to value the output as a weighted average of import replacement and export production, where the proportion used for import replacement will usually increase over time as the size of the domestic market grows.

The direct export of gas, generally as LNG, calls for a slightly different approach to benefit estimation. The value of the product is the price received from the importing country, less the capital and operating costs of the regasification and transport facilities required. This calculation will yield an export-equivalent price at the border, from which the capital and operating costs of liquefaction must be subtracted to arrive at total benefits. To calculate the netback value of LNG it is important to take account of the volumetric losses associated with transport and regasification in the denominator of the equation. Chapter 10 illustrates this procedure in detail.

6.3 Summary

The allocation of gas among competing uses, or the formulation of a gas pricing policy, relies heavily upon the proper measurement of gas benefits. Yet those figures, in turn, must reflect the particular technical and economic circumstances of the project and country in which the gas will be used. In practice, there is wide variation in the value of gas both across projects in the same country and across countries for the same kind of project. The key point to remember in gas benefit valuation is that the benefits must be assessed relative to the next-best alternative for obtaining the end product.

There is no one best measure that summarizes gas benefits. The netback is a useful figure in comparing marginal uses of gas that are roughly similar in size and which exhibit roughly similar economies of scale. It is also defined in a consistent way with the average incremental cost of gas, thereby permitting preliminary judgements on project profitability to be made. However, it cannot substitute for a full gas planning model (see Chapter 12) to rank dissimilar options, and it is a poor yardstick for pricing in most cases (see Chapter 10).

The remaining chapters of this part illustrate these benefit measures in detail, and introduce variations where appropriate. They also provide important insights into the sensitivity of gas benefits to particular technical and economic factors that are relevant for the major gas-using sectors. We begin with LNG export because it provides the most straightforward application of the netback concept.

7 The Value of Gas in LNG Exports

Although the export of liquefied natural gas is a realistic possibility for only a small number of countries, it provides a good initial illustration of how to measure gas benefits and the relationship between the netback value of gas and the net present value of the project as a whole. In addition, in many developing countries, the first question that is asked following the discovery of gas is whether LNG export is feasible. This chapter aims to clarify some of the economic issues that need to be addressed in answering that question. It does not consider, however, the additional financial and legal problems whose resolution is equally crucial for the success of an LNG project.

After a review of the background and prospects of LNG trade, we describe the structure of an LNG project and develop three project prototypes. We then explore a range of seven cases based on these prototypes in order to identify the major technical parameters that affect LNG netbacks.

7.1 The Evolution and Prospects of LNG Trade

International trade in LNG began with the trial shipments from Louisiana to Canvey Island (UK) in 1954.[1] Its success led to the first commercial base-load international LNG project in 1964 between Arzew, Algeria, and the UK for about 40 bcf per annum over a fifteen-year contract period. This was followed by ventures between Algeria and France in 1965, and Alaska and Japan in 1969. Gas exports grew about tenfold between 1966 and 1980 because of the mutual benefits to exporters and importers of LNG. For exporting countries, gas that would otherwise have been flared (in Abu Dhabi and

---

[1] The use of LNG for peak shaving began on a small scale in the US in the early 1940s.

Libya), or gas that was surplus to long-term domestic needs, could be exported as LNG to generate foreign exchange. LNG projects provided an important option for developing countries with relatively abundant unutilized natural gas reserves.

LNG projects have complex structures, however, and are very difficult to arrange. Projects may involve several parties at each stage. For example, in Indonesia at the production stage, the owner (Pertamina) has 100 per cent control and works with the operator (Mobil Oil Indonesia) under a production-sharing contract. Pertamina also owns the liquefaction plant, and three operators are responsible for processing (PT Arun Natural Gas Liquefaction Company, Mobil Oil Indonesia and Japan Indonesia LNG Company). Burmah Gas Transport Limited is the contractor for shipping operations under a transportation agreement with Pertamina. The latter is responsible for the sale of gas to end-users, including Nippon Steel Corporation, Osaka Gas and three Japanese electric power companies, all under a single contract.

The pattern of involvement of the parties in LNG projects differs and it often changes over time. For example, prior to 1971 in Algeria, the government and the private sector companies had a mixed equity participation in LNG projects. Since 1971, the government has increased its share to 100 per cent of the total equity in the project up to the f.o.b. point and to a 50 per cent share in shipping. In Brunei, the government had a 10 per cent minority share in the liquefaction plant only and later extended this to a one-third share in the project, including the shipping and trading.

Before 1973, prices of gas for importing countries were low relative to the prices of alternative energy sources. Between 1973 and 1979, LNG prices remained competitive, though they were increasingly linked to the prices of petroleum products. In countries with a serious pollution problem, such as Japan, and more recently South Korea, LNG also had a premium value as a clean fuel. Several European countries had gas pipeline networks (to distribute town gas produced from coal) which could be used to distribute natural gas. In some instances, LNG imports were needed to maintain the supply of gas to existing distribution networks, where not enough gas was available locally. For other importers, LNG provided an economic way of diversifying the sources and types of energy, to improve the overall security of supply. For both buyers and sellers, LNG became a proven means of supply which was technically reliable and safe, and also offered the most economic means of bringing large volumes of gas to markets where delivery by pipeline was impractical.

Much larger LNG projects were planned in the 1970s to exploit economies of scale in liquefaction and to meet increasing demand. The first of this new generation of large-scale projects was the export of gas from Brunei to Japan which started in 1972. World LNG trade increased from 2.7 bcm in 1970 to over 60 bcm in 1988. The share of LNG in total gas trade increased from about 6 per cent to over 23 per cent over the same period. Table 7.1 shows LNG exports in 1988 by country of origin.

Except for the implementation of projects already under construction, the growth in international LNG trade virtually stopped in the early 1980s. Some of the projects proposed in the 1970s, such as the export of gas from Iran to the US and Japan, and from Algeria and Nigeria to the US and Western Europe, were either cancelled or put on hold. Actual trade remained nearly static at the 1980 level of around 1.5 tcf per annum. This sluggishness was generally attributed to the economic recession of the early 1980s, the fall in oil prices during the mid-1980s, widespread energy conservation, and a switch from energy-intensive industries to the manufacture of

Table 7.1: LNG Exports by Country of Origin. 1988. Billion Cubic Metres.

Exporter	Volume
Abu Dhabi	3.18
Algeria	14.83
Brunei	7.25
Indonesia	24.58
Libya	1.06
Malaysia	8.26
USA	1.30
Total	60.46

Source: *Le Gaz naturel dans le monde en 1988*, Cedigaz, 1989.

less energy-intensive products and services in developed countries. In 1987, LNG exports increased by about 10 per cent, largely owing to the increase in Algerian gas deliveries, which had been pending renegotiation of contracts, as well as the additional exports from Indonesia to South Korea. More recently, with the increased competitiveness of gas prices, and the willingness of exporters to increase revenues from gas exports, as well as environmental concerns in many importing countries, the interest in LNG projects has picked up. A number of countries are working on the expansion of existing projects or starting up projects that had previously been put on hold. LNG trade is expected to resume its growth in the 1990s, but this may be at a slower rate than in the 1970s.

Developing countries are responsible for almost all LNG exports (with the exception of small US deliveries to Japan), and are likely to remain the major suppliers of LNG. The Middle East holds over 25 per cent of the world's proven gas reserves, and more than half of these are in Iran. Within this region, only Abu Dhabi has an operating LNG project while Iran, Qatar and the UAE have a large export potential. In Africa, Algeria and Libya are already LNG exporters and other possible future exporters include Nigeria as well as Cameroon, Egypt and Angola. In Latin America there are no current LNG projects. There are, however, pipeline exports from Mexico, which has the largest gas reserves in the region, to the USA, and from Bolivia to Argentina. There are large gas reserves in this region and there is a growing regional market; Bolivia and Brazil are studying a large pipeline project and the Mexican–US trade is expected to grow. In Asia, Malaysia holds the largest gas reserves, and together with Indonesia and Brunei has operating LNG projects. In this region Bangladesh and possibly Thailand and Burma could emerge as LNG exporters. A few developed countries may also start exporting LNG.

Most of the demand for LNG will continue to come from developed countries. As discussed in Chapter 3 above, the three major gas-consuming areas are the USA, Western Europe and Japan. The USA is again considering additional LNG imports from a number of sources including Nigeria and Norway.

Western Europe has a high level of domestic production, and already imports LNG and pipeline gas from North Africa and the USSR. In particular, since the agreement with Algeria about gas pricing, potential new LNG projects are expected to come on stream.

Japan is expected to remain the largest importer of LNG at least until the year 2000. Its dense urban concentrations require tight pollution control, which puts a premium value on the clean-burning characteristics of gas. Japan currently imports LNG from Abu Dhabi, Alaska, Brunei, Indonesia and Malaysia, and in the future may also import LNG from Australia and Canada. The major users of LNG in Japan will remain the power utilities, which currently account for 75 per cent of total gas use, followed by industrial and residential users.

Since 1986, the contractual terms for gas exports have changed dramatically. Prior to 1986, the price of natural gas was closely linked to the prices of crude oil or specific petroleum products. Since 1986, gas prices have been linked to the prices of competing sources of energy so that gas remains competitive at the consumer level. Therefore in many new markets gas prices have been renegotiated and gas prices, which reached their highest levels in the early 1980s (when they rose above $6.60/mcf c.i.f. ex-regasification in some markets), fell to their lowest level in early 1987 (see Table 7.2).

7.2 The Structure of LNG Projects

LNG projects require large proven reserves and take about a

Table 7.2: Evolution of Gas Prices, c.i.f. before Regasification.[a] 1983–8. Constant 1982 Dollars per Million British Thermal Units.

	1983	1984	1985	1986	1987	1988
US	4.3–5.5	4.3–5.1	3.3–4.4	2.6	2.3	1.9
Western Europe	3.5–4.4	3.5–4.2	3.4–4.4	3.2–3.6	2.5–2.8	1.9–2.4
Japan	4.9–5.3	4.8–5.2	4.7–5.2	3.2–4.7	3.2–3.5	3.1–3.8

Note: (a) All prices are quoted for mid-year.
Source: *Le Gaz naturel dans le monde en 1988*, Cedigaz, 1989.

decade between the first indication of interest and their commissioning. They are highly capital intensive, and require considerable up-front investments. In addition they usually require extensive investments in ports, housing for employees and community-related services. A study of an LNG project can cost tens of millions of dollars. The costs of a typical liquefaction plant, tankers and receiving terminals amount to $1.4–1.7 billion for a 300 mmcf/d project and $2.0–2.5 billion for a 500 mmcf/d project.

An LNG export project consists of four distinct but interrelated stages:

(a) Gas production, treatment and transport to the liquefaction plant;
(b) Liquefaction, storage and ship loading;
(c) Shipping LNG in specially cooled (cryogenic) tankers to the reception terminal;
(d) Arrival at receiving terminal, unloading, LNG storage and regasification.

Although these activities are generally conducted by separate entities, an LNG project requires high degrees of interdependence and interaction between suppliers and customers (see Figure 7.1). Each phase of an LNG project is part of an integrated system stretching from the gas well to the ultimate consumer. If any one element in the chain is not ready in time or fails for any reason, the whole project may be in jeopardy.

The project must also operate at a high load factor and over a contract period from fifteen to twenty-five years in order to justify the enormous investments required. Neither the supplier nor the customers can easily turn elsewhere for outlets or alternative supplies of LNG of the magnitude involved. LNG trade differs from oil trade in that the opportunities for spot cargo trading of base-load LNG for conventional uses are small. There is, however, some opportunity for spot cargo trading for peak-shaving purposes.

Export volumes tend to be large, and the reserves dedicated to an export scheme should be sufficient to sustain production over the whole of the contract period, with a safety margin of about 30 per cent, particularly if part of the supply comes

The Value of Gas in LNG Exports 79

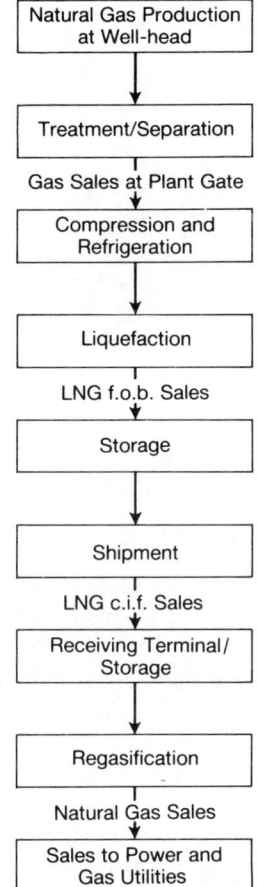

Figure 7.1: An LNG Export Scheme.

from associated gas. LNG projects require very reliable estimates of reserves and production. Reserves of about 3 tcf provide a sufficient threshold for LNG projects with an approximate capacity of 300 mmcf/d; projects based on recoverable reserves of 4–5 tcf benefit from economies of scale.

The gas that enters the liquefaction plant is treated and free of most impurities. If the gas contains a high percentage of carbon dioxide, hydrogen sulphide, nitrogen or metal particles, additional investment in pre-treatment equipment is required to reduce the risk of disruption to the production

80 *The Economics of Natural Gas*

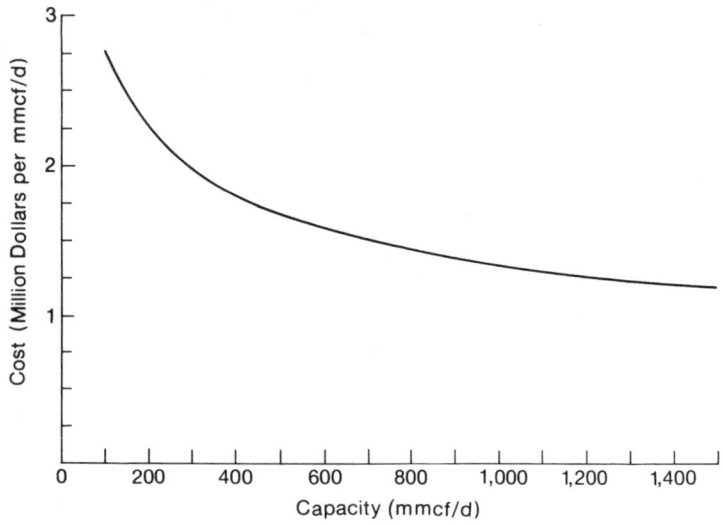

Figure 7.2: Liquefaction Plant Cost.

process. In the calculations that follow, the gross thermal content of gas is assumed to be 1,000 btu/scf. This is a very lean gas stream. In practice a higher or lower calorific value can significantly affect the costs and benefits of the project.[2] Once gas enters the liquefaction plant, it is cooled to about −161 °C at atmospheric pressure to reduce its volume to 1/600th of the equivalent quantity of gaseous methane. Each process train or liquefaction unit generally involves further purification and dehydration of the incoming gas followed by compression, refrigeration and liquefaction. The liquefaction plant is often the most expensive link in the LNG chain. For example, a 500 mmcf/d liquefaction plant costs about $800 million, excluding the costs of site prparation. The most costly items within the plant are the liquefaction trains, and steam and power generating facilities, each accounting for 25 per cent or more of the total cost.

---

[2] A rich gas with a high content of heavy hydrocarbons can be marketed in two ways. First, the heavy gases can be extracted, fractionated and transported by LPG carriers. This will increase the net present value of the project's cash flow and also the netback. Second, the heavier gases can be liquefied with the methane and transported to the importing country together. For some buyers this gas will have a premium value. Mixing of rich and lean gas may cause problems for other end-users and require adjustment of burners.

The 500 mmcf/d plant size was chosen for the calculations below to represent a standard-size liquefaction plant. The possibility of scaling down the liquefaction plant was investigated. The results indicate that for a plant size below 300 mmcf/d, costs begin to rise quite rapidly (see Figure 7.2). Table 7.3 shows the major cost items for the 500 mmcf/d and 300 mmcf/d liquefaction plants.

Once liquefied, gas is moved into tankers for transport. The current standard for new tanker designs is 130,000 cubic metres. This standard has evolved as a compromise between the desire to take advantage of economies of scale and the need for LNG carriers to be able to enter European, Japanese and American ports.[3] The capital costs of the two major tanker designs, the membrane and spherical types, depend on the shipyard, but a price of $150 million for a 130,000 cubic metre tanker is often quoted for both designs.[4]

The operating costs cover fuel, harbour fees, maintenance, insurance and wages. Operating costs are sensitive to distance. For a typical route of 5,000 nautical miles one way, which corresponds approximately to a North African–European project, operating costs constitute 7–8 per cent of capital costs. The study described below assumes zero boil-off since current tanker design has reduced the maximum boil-off of gas *en route* to 0.11 per cent per day at the highest.[5] Improved insulation of new carriers and the installation of a small reliquefaction system also reduce boil-off.

The number of LNG tankers that are needed for a project depends on the distance between ports, the speed of the tankers, tank filling rates, expected downtime, the desired safety margin and other system characteristics, as well as the overall volume of LNG transported. The prototype projects we simulated took account of these factors and require five vessels for

---

[3] An additional factor in the design of modern tankers is safety regulations. The US Coast Guard regulations set global standards.

[4] There are more than sixty LNG tankers in the world. The membrane design had an early lead, but recently the spherical design seems to have taken over because of its greater flexibility in quantities carried and the ease of inspection of the spherical tank.

[5] The boil-off is the gas vaporized by the leakage of heat into the LNG tank. In conventional LNG carriers, the LNG boil-off is used to supply a portion of their fuel needs.

Table 7.3: Liquefaction Plant Cost Breakdown. Million 1982 Dollars.

	500 mmcf/d	300 mmcf/d
Liquefaction Section	290	170
Utilities and Auxiliaries	330	230
Site Preparation and Building	35	30
Storage	90	70
Marine and Loading	110	95
Total	855	595

Table 7.4: Cost Breakdown at Receiving Terminal under Various Scenarios. Million 1982 Dollars.

	Scenario A 500 mmcf/d	Scenario B 300 mmcf/d	Scenario C Multiple destination[a]
Regasification	125	75	90
Utilities and Auxiliaries	90	55	70
Site Preparation and Building	20	15	60
Storage	100	80	240
Marine and Loading	130	120	370
Total	465	345	830

Note: (a) These figures include costs at regasification plants/receiving terminals at all three destinations.

the medium-size, single-destination project (Scenario A), three vessels for the small-size, single-destination project (Scenario B), and one vessel for the small-size, multi-destination project (Scenario C).

The receiving terminal and regasification plant is the simplest and least expensive phase of the LNG chain. It consists of a harbour with facilities for offloading tankers, LNG storage, regasification and distribution of gas. The docking facilities typically cost at least $120 million and do not vary much with the size of the terminal. The port and storage facilities may represent over half the total costs of the receiving and regasification terminal. Table 7.4 shows a breakdown of these costs.

The Value of Gas in LNG Exports 83

There are two principal processes for regasification. In large installations, sea water is used to raise the LNG temperature. For small and peak-shaving plants, gas burners are used.[6]

7.3 Simulation Study of Natural Gas Benefits

To estimate the netback value of gas in LNG export, a simulation study was carried out based on data from a number of actual projects and several feasibility studies for future projects.[7] Seven cases were developed in all to identify the major technical and economic parameters that affect system costs and netback values. These cases were derived from the three basic project scenarios listed in the previous section and cited in Table 7.4.

A real discount rate of 10 per cent is assumed in all cases, and the other key assumptions are as shown in Table 7.5. Scenario A is an average-size LNG project with a capacity of 500 mmcf/d situated some 5,000 miles from the export market (e.g. a project to export from North Africa to Europe) and served by five LNG tankers. Its Case I (with infrastructure available) is used as the base case for sensitivity testing. Case II assumes that no infrastructure is available at the site so that its costs form part of the capital costs of the project. Case III tests the effects of a change to a lower gas price of $3.00/mcf.

Scenario B is a smaller version of Scenario A with a capacity of 300 mmcf/d and three tankers going to a single destination. This is probably close to the minimum economic size for an LNG project. Three variations of Scenario B are considered: Case IV with infrastructure available; Case V without; and Case VI with infrastructure available but with a lower price.

Scenario C – or Case VII – is a multi-destination project with three small markets (perhaps on a long coastline of a

---

[6] The capital cost of the gas vaporization process is only half that of using sea water, but this difference tends to be offset by the cost of the gas used in the gas burners.

[7] For further details on the capital and operating cost schedule, volume of gas, and prices used in all simulations see Mashayekhi, A. and Jensen Associates, Inc., *LNG Export Opportunities for Developing Countries and The Economic Value of Natural Gas in LNG Export*, Energy Department Paper No. 12, World Bank, December 1983.

Table 7.5: Assumptions for Seven LNG Simulations.

Scenario	Case	Volume[a]	Infrastructure Available[b]	Transportation Distance[c]	Number of Receiving Terminals
A	I	500	Yes	5,000	1
	II	500	No	5,000	1
	III[d]	500	Yes	5,000	1
B	IV	300	Yes	5,000	1
	V	300	No	5,000	1
	VI[d]	300	Yes	5,000	1
C	VII	300	Yes	1,000	3

Notes: (a) Liquefaction plant input, million cubic feet per day.
(b) At liquefaction plant site.
(c) Nautical miles, one way.
(d) In cases III and VI, the gas price is assumed to be $3.00/mcf, and in the other cases it is assumed to be $4.44/mcf.

single country) located about 1,000 miles from the gas liquefaction plant. It is a barebones project with a single LNG tanker. Downtime is allowed for tanker maintenance.

7.4 The Netback Calculation

The netback value depends not only on the price received for the LNG and the cost of delivering it to the user, but also on the volumes delivered. To estimate these it is necessary to account for losses – or burn-off – at each stage of the LNG project. Because costs at each stage are also discrete, it is possible to estimate the netback value of the gas as it moves from one stage to the next.

As an example of a netback calculation, one of the simulations based on Scenario A (Case III) assumes that the value of gas at the point of entry into the importing country's main transmission line *ex-vaporization* is determined by the delivered price and quantity of gas over the twenty years of the project's operation. In this example that value is $3.00/mcf.

The netback ex-ship is estimated by deducting the present value of the capital and operating costs for the receiving terminal, storage and regasification facilities, which is $433 million, from the present value of the total revenue from the sale of gas and dividing the result by the present value of the gas volume delivered into the main transmission pipeline of the importing country. This results in a netback value of $2.59/mcf at the delivery point into the receiving terminal. Algebraically,

$$N_{\text{ex-ship}} = \frac{\sum_{t=1}^{20}[PV_1/(1+r)^t] - \sum_{t=-5}^{20}[(C_1+O_1)/(1+r)^t]}{\sum_{t=1}^{20}[V_1/(1+r)^t]} \quad (7.1)$$

where P is the price of natural gas delivered into the importing country's pipeline, V_1 is the volume of gas delivered each year, C_1 and O_1 are respectively the capital and operating costs related to regasification, storage and the receiving terminal, and r is the discount rate.

The netback ex-liquefaction is estimated similarly by deducting the present value of capital and operating costs of all receiving, storage, regasification and shipping facilities from the present value of total revenues from the sale of gas and dividing by the present value of the gas volume transferred into the ship. Algebraically,

$$N_{\text{ex-liquefaction}} = \frac{\sum_{t=1}^{20}[PV_1/(1+r)^t] - \sum_{t=-5}^{20}[(C_2+O_2)/(1+r)^t]}{\sum_{t=1}^{20}[V_2/(1+r)^t]} \quad (7.2)$$

where PV_1 is the total revenue from gas sales, $C_2 + O_2$ is equal to $C_1 + O_1$ plus the capital and operating costs of shipping, and V_2 is the volume of gas transferred into the ship.

The netback at the point of entry into the liquefaction plant (ex-pipeline) is estimated by deducting the capital and operating costs of the liquefaction and storage facilities from the present value of the revenues from the sale of gas, and dividing by the discounted volumes of gas delivered to the liquefaction plant. Algebraically,

$$N_{\text{ex-pipeline}} = \frac{\sum_{t=1}^{20}[PV_1/(1+r)^t] - \sum_{t=-5}^{20}[(C_3+O_3)/(1+r)^t]}{\sum_{t=1}^{20}[V_3/(1+r)^t]} \quad (7.3)$$

86   *The Economics of Natural Gas*

where $C_3 + O_3$ is equal to $C_2 + O_2$ plus the liquefaction and storage capital and operating costs, and V_3 is the volume of gas delivered to the liquefaction plant. The netback value at this point is $1.00/mcf.

7.5 Results of the Simulations

The results of the netback calculations for the cases considered are summarized in Table 7.6. For Case I, the netback value of gas at the point of entry into the liquefaction plant is $2.43/mcf. This is based on a discount rate of 10 per cent and a price of gas at the point of entry into the receiving country (i.e. after shipping and regasification) of $4.44/mcf.

If infrastructure is needed at the liquefaction site (Case II), then the netback falls to $2.06/mcf. The netback value is very sensitive, however, to the price received in the importing country. Case III assumes that the low 1986 prices will continue over the twenty years of a given LNG project, in which case the netback falls to $1.00/mcf.

The second scenario is centred on a liquefaction plant of 300 mmcf/d, to examine the diseconomies of scale. Because of the different production schedule, the present value of the gas stream entering the gas grid in the importing country in Case IV, for example, is somewhat lower than in Case I, at $4.37/mcf. Working back through the LNG chain, $0.46/mcf is netted out in the regasification plant, which is 13 per cent

Table 7.6: Assumed Price and Unit Netback Values. 1982 Dollars per Thousand Cubic Feet.

Scenario	Case	Price at Regasification in Importing Country	Netback at Entry into Liquefaction Plant
A	I	4.44	2.43
	II	4.44	2.06
	III	3.00	1.00
B	IV	4.37	2.25
	V	4.37	1.86
	VI	2.93	0.81
C	VII	4.37	2.00

more than in Case I. Shipping costs are similar since LNG carriers are used with roughly equal efficiency in both cases. Unit liquefaction costs rise from $0.99 to $1.07 per mcf, although the expected diseconomies of scale are moderated by a shorter construction period and a faster build-up of production. All things taken together, there is surprisingly little difference between the netbacks in Cases I and IV. Diseconomies of scale reduce the netback by less than 10 per cent to $2.25/mcf. Limited infrastructure, and therefore higher capital costs at the liquefaction plant site, reduce the ex-pipeline netback to $1.86/mcf. With a lower gas price of $3.00/mcf the netback falls to $0.81/mcf.

Scenario C illustrates the economics of a small-scale multi-destination LNG system where the gas is carried over a short distance of 1,000 miles to several small markets. The major difference from the other cases is that regasification costs are higher. Even after substituting gas-fired vaporization, which is more appropriate to smaller systems, the regasification cost is close to three times as large as in Case I. Offsetting this somewhat are lower costs in the transportation phase. The short hauls in Scenario C bring down the fuel cost dramatically. Since the cost per day in harbour is far below the operating cost at sea, the overall effect is that the transportation cost is about one-third that of Case I. The unit netback ex-pipeline in this case is $2.00/mcf compared with $2.43/mcf in Case I.

It should be remembered that these netback figures relate to the value of the gas as it enters the liquefaction plant. To assess the viability of the entire LNG system (or to estimate the value of gas at the well-head), one would have to subtract the 'upstream' costs of exploration, production and delivery to the liquefaction plant. Alternatively, one could calculate the net present value (NPV) of the entire LNG system, taking into account the cost of delivering gas to the plant.

The results of the latter calculation for Cases I, II and III are shown in Table 7.7. We have taken three alternative values for the cost of gas delivered to the liquefaction plant: zero (for reference, or as though the plant were based on associated gas which would otherwise be flared); $0.50/mcf (a low value based on the empirical results in Chapter 4); and $1.00/mcf (a moderately high value). The resulting project

Table 7.7: Net Present Values of Projects under Different Gas Cost Assumptions for Scenario A. Million 1982 Dollars.

Case:	I	II	III
Gas Costs[a]			
$0.00/mcf	2,284	1,935	969
$0.50/mcf	1,814	1,435	499
$1.00/mcf	1,343	994	29

Note: (a) Cost of gas delivered to the liquefaction plant.

NPVs are highly sensitive to the assumed gas cost. The three cases achieve economic viability in the sense of having an NPV greater than zero, but at a gas cost of $1.00/mcf the project's NPV is 25–45 per cent lower than with a gas cost of $0.50/mcf. Further, with a low gas price of $3.00/mcf exvaporization (over the life of the project), the project's NPV is quite low once the cost of gas increases to $1.00/mcf.

7.6 Conclusions

Under the right circumstances LNG projects can be an economically attractive use of gas. However, because of their large investment requirements and their long-term, inflexible nature, they are also subject to major risks on both the supply and demand sides. These risks – largely stemming from the inherent uncertainties of the energy market – should be examined in a consistent quantitative framework such as that described in this chapter.

On the supply side, we have reached the somewhat surprising conclusion that smaller-size projects, and even small multi-destination projects, do not have dramatically higher unit costs than medium-size projects, which implies that there may be more scope than was previously thought to reduce project risks by starting with a smaller investment or by serving nearby, smaller markets. It is also clear, however, that an LNG project of whatever size is unlikely to make sense unless the cost of gas (including exploration, production and delivery to the liquefaction plant) is below about $1.00/mcf.

On the demand side, prospects for new LNG projects are

not expected to begin to improve in the 1990s. The major importing markets in the US and Western Europe have access to competing supplies via pipeline; none the less, the expansion of some existing projects is currently under study. Japan and South Korea are also looking at potential additions to their LNG supplies from Asia and the Middle East.

Given the experience of the mid-1980s, fluctuations in oil prices must be counted as an additional major risk that must generally be borne by the exporting country. The results reported in this chapter indicate the sensitivity of netbacks to gas prices.

For all these reasons, the LNG export market should be very competitive until the mid-1990s, and only a few developing countries are expected to benefit from the LNG option. In any case, its netbacks and NPVs are lower than those of most domestic uses of gas, as examined in the next two chapters. However, countries with large reserves of surplus gas over a long period are likely to find the export of LNG an economically attractive investment.

8 The Value of Gas in Power and Industry

For many of the fifty or so developing countries with natural gas resources, the power and industrial sectors represent the largest potential gas users. The early use of natural gas for power generation can provide significant benefits to both the power and gas sectors. For a power utility facing high fuel bills and perhaps energy shortages, gas provides a rapid and economic source of energy. At the same time, to ensure financial viability in the gas sector, the large initial investments in gas infrastructure must be recouped by major early gas sales. In many developing countries with limited industrial and residential energy demand, the power sector can provide this initial market. Over the longer term, if there is a gas supply constraint and other higher-value markets develop, they will gradually displace gas-based power generation from the base load and provide the early financial returns that are necessary to justify the initial infrastructural investment.

The industrial sector is often a major gas consumer. It is difficult to generalize about the value of gas in industrial uses because the range of uses within the sector is a broad one, encompassing both fuel and feedstock applications. In this chapter we consider four types of industrial gas use, which account for the bulk of this demand: fuel use, ammonia feedstock, methanol production and petrochemical feedstock. Our focus is on the variety of possible gas uses in this sector, and the factors that will influence the value of gas in each, rather than on numerical results which will vary significantly with market conditions.

8.1 The Value of Gas in Power Generation

The determination of the economic value of gas used to produce electricity must generally be based on the configuration of the power system as a whole, not merely on the gas-using plant. This is because converting a plant from oil to gas, or adding a new gas-based plant to the system, will often change the usage patterns of other generating plants as the system planners minimize total system costs based on the new configuration. This chapter reports the results of a simulation study undertaken to evaluate the economic value of natural gas in power generation using the netback approach, using data for a medium-sized Asian country which has endowments of hydroelectric and coal resources as alternatives to gas.[1]

In order to keep the exposition as simple as possible, we do not consider the effects that a reduction in prices due to the use of gas may have on the demand for electricity. If the substitution of gas for, say, fuel oil significantly reduces the cost of power generation, and if those savings are passed on to the consumer, then the demand for gas-based electricity may be expected to grow faster than the demand for oil-based electricity would have grown.

The netback value of gas for power generation is defined as the cost savings for the power system as a whole that would accrue if gas were substituted for other fuels. In order to derive this, it is necessary to develop least-cost power system expansion plans with and without given quantities of gas, and then to compare the present values of the total costs for each plan.[2] When the total savings in discounted costs are divided by the discounted gas consumption, the result is the netback value of gas at the point of delivery to the project. Gas supply costs have not been included, so the netback value of gas also represents the 'break-even' price at which the costs and benefits of using gas in power generation are equalized over the life of the calculation.

This netback value is a function both of the inherited plant mix (hydro, oil-fired, coal-fired, gas turbine, etc.) and of optimal future mixes of plants and fuels. The optimal future

---

[1] For full details of the study and its assumptions see Albouy, Y. and Mashayekhi, A., *Value of Natural Gas in Power Generation*, Energy Department Paper No. 19, World Bank, November 1984.

[2] We do not discuss the techniques of power system planning. For a review of these and their links to marginal costs see Albouy, Y., *Marginal Cost Analysis and Pricing of Water and Electric Power*, Inter-American Development Bank, Washington DC, 1983.

mix will itself depend on the capital costs of new plants, the costs of converting inherited plants to burn natural gas, the quantities of gas available, estimates of the future prices of gas and other fuels, and assumptions about rates of growth of demand and the discount rate. It will also be affected by any differences in the availabilities of gas-burning and other types of thermal plant, differences in annual non-fuel variable costs and maintenance costs.

We first optimize the power system expansion plan without gas, then reoptimize the system with a small supply of gas that would be burnt in new gas-fired plants, and finally repeat this step for progressively larger supplies of gas that may be used both in new gas-fired plants and in existing plants that are converted to gas firing. Finally, we estimate the benefits to be derived by constructing capacity ahead of need in order to speed up the rate of utilization of gas.

The country considered has a medium-sized power system, with a maximum demand of 2,400 MW and a load factor of 70 per cent. It has a sizeable amount of peak-shaving hydro capacity, and thermal generation comes from existing or committed lignite-fired plants, steam plants on residual fuel oil, gas turbines and diesel units. The assumed annual rate of load growth is 7 per cent.

The generating system is simulated over a period of twenty years, with an adjustment made for benefits and costs incurred beyond that period. The conversion of existing oil-fired steam plants and committed gas turbines is possible, and we assume that the projected plants in the country under study have economic lives of twenty years for steam plants and fifteen years for both gas turbines and combined-cycle plants.

8.2 Netback Values

The average netback values of gas for the country, as reported in Table 8.1, are based on the conversion of inherited steam capacity and the construction of new gas-fired combined-cycle units. When allowance is made for conversion, the value of gas declines continuously as the supply of gas is increased.

The discussion so far has been based on *average* netback values. The fact that these decrease as more gas becomes

Table 8.1: Average Netback Value[a] of Gas in the Power Sector in Example Country. Volumes in Million Cubic Feet. Values in 1982 Dollars per Thousand Cubic Feet.

Gas Available	Gas Used in Final Year	Average Value
20	16.6	4.17
60	45.6	4.14
87	66.1	3.90
300	228.0	3.67
1,800	1,131.0	3.39

Note: (a) These calculations assume a discount rate of 10 per cent.

Table 8.2: Incremental Netback Value[a] of Gas in the Power Sector in Example Country. Volumes in Million Cubic Feet. Values in 1982 Dollars per Thousand Cubic Feet.

Gas Available	Gas Used in Final Year	Incremental Value
20	16.6	–
60	45.6	4.13
87	66.1	3.36
300	228.0	3.55
1,800	1,131.0	3.20

Note: (a) These calculations assume a discount rate of 10 per cent.

available indicates that the marginal, or *incremental*, netbacks must be lower than the *average* netbacks. The expansion in gas use should be stopped at the point where this *incremental* netback falls below the cost of gas. It is the incremental netback – not the average – that indicates the willingness to pay for gas in the power sector.

Table 8.2 shows that the incremental values of gas as the supply is increased are significantly lower than the average netbacks shown in Table 8.1 and that they, too, fall over the gas supply range shown. Additional increments to gas supply beyond 1,800 mmcf/d lead to an incremental netback of zero.

The netback is higher than the price of the fuel in all the cases considered owing to savings in capacity costs. The netback on total gas consumption is generally higher than that for incremental consumption at the margin. However, the lumpiness of investment produces some unexpected results. For example, a higher growth rate of supply generally increases both peak demand and the netback.

A useful preliminary step might be to postulate a profile of gas prices and to derive the gas demand from them with the help of least-cost planning methods. This can provide an initial picture of the pattern of lumpiness of gas demand which, in turn, can be used to select the quantity increments of gas supply for the netback calculation.

8.3 The Value of Gas in Industrial and Chemical Uses

The first main application of gas in the industrial sector is as a fuel for steam raising, metalworking, process heat for cement production, and agricultural product drying. Gas competes with coal and fuel oil in these uses, and with diesel oil in small industries and those where the quality of the heat is important, as in glass production.

Secondly, gas is used as a feedstock in the production of ammonia, which is used in turn to produce urea and other nitrogenous fertilizers. Alternative feedstocks include coal, naphtha and fuel oil.

The third main use of gas in the industrial sector is in the production of methanol, although this use is not as common as the first two. Methanol is an important chemical intermediate good and has potential as a gasoline substitute. Its traditional chemical applications include the production of formaldehyde, which is used in turn to produce a variety of resins and adhesives. Methanol is also used as a solvent and in the production of acetic acid. Naphtha and fuel oil are the alternative feedstocks currently employed, and production from coal is considered technically feasible although it is not yet in commercial use.

Finally, the ethane portion of natural gas can be used as a feedstock for the petrochemical industry. The gas is fed into a steam cracker, and a range of products is obtained, including ethylene and propylene. These are the basic building blocks in the production of polyethylene, polypropylene and other intermediate inputs, which are subsequently processed into a variety of petrochemical final products, from plastic containers to vehicle antifreeze. Petroleum-derived products such as naphtha and gas oil are alternative feedstocks for the steam cracker.

As discussed in Chapter 6 above, in estimating the value of gas in specific industrial applications, it is important to identify the most economic alternative course of action in the absence of gas. These alternatives include: (a) producing the same final product domestically using some alternative fuel or feedstock; (b) importing the final product; and (c) using some alternative final product (e.g. jute bags instead of polythene). For example, in the case of the conversion of a cement plant boiler from coal to gas, the netback value of gas would equal the present value of the cost of coal saved less any capital and operating costs incurred in the conversion of the plant, divided by the present value of the volume of gas used. In the case of an existing urea plant, the netback should be calculated from the cost of the alternative feedstock saved (e.g. naphtha), less any costs of conversion, and in the case of a new urea plant, from the cost of imported fertilizer less the capital and operating costs of the domestic fertilizer plant. In the remainder of this chapter we discuss the principal industrial uses of gas in greater detail and discuss the technical features that are likely to influence the netback value of gas in each use.

(a) Gas as an Industrial Fuel. Natural gas offers a number of advantages over the various alternative fuels used in industry. Gas burns more cleanly than other fuels. Any hydrogen sulphide in the raw gas is generally removed at the gas processing plant, while the hydrocarbons in gas are burnt completely, unlike those in fuel oil or coal. This is an important benefit, especially in areas of industrial and population concentration where pollution is a major concern, and in industries where it is critical to avoid the contamination of products. Such considerations have led Japan to place a major emphasis on the use of natural gas as a fuel.

In addition to these clean-burning qualities, the temperature of a gas flame is more easily controlled than that of a coal or even a liquid fuel flame. This is a crucial consideration in certain process heat applications. Gas can also provide heat at a higher temperature than can conventionally be provided by coal or oil.

In assessing the value of gas in fuel uses, the principal elements will be the economic cost of the alternative fuel substituted, such as coal or fuel oil, and any differential capital and operating costs. This value is highly location specific, depending, for example, on the costs of coal production and transportation to the particular site. In a country the size of China, considerable variation in value from one location to another can be expected, and thus evaluation of the benefits of gas should be made on a location-specific basis. Given the low oil prices of the late 1980s, netbacks to gas in fuel uses vary in a range of $1–4 per mmbtu depending on the price of the fuel substituted (coal, fuel oil or diesel oil) and conversion costs.

In general, gas will have a higher value in new industrial facilities than when an existing plant is converted from coal to gas. This is because the use of gas avoids many of the plant costs that are incurred in coal use when a facility is being designed and built (such as the provision of handling and storage facilities, a coal pulverizer, cleaner and drier, and equipment for ash removal and pollution control). In the case of conversion from coal to gas, these costs will already have been incurred.

(*b*) *Gas as an Ammonia Feedstock.* Natural gas is the most commonly-used feedstock for the production of ammonia, which is the base for nearly all nitrogenous fertilizers. Of the total world consumption of fertilizers, nitrogen represents about 50 per cent of the nutrients. Over 70 per cent of world ammonia production capacity is based on natural gas, and if smaller Chinese plants are excluded from the total, about 80 per cent of this capacity is based on gas. Ammonia is also the base for two other groups of products: (a) explosives; and (b) fibres, plastics and resins.

World consumption of nitrogenous fertilizers grew at an average annual rate of 6.6 per cent between 1970 and 1980. Growth was slowest in the OECD countries, where the levels of fertilizer use were already close to optimal. In the rest of the world, where there was scope to increase fertilizer use, growth was more rapid. The Chinese growth rate was higher than that of any other region, as consumption grew at an average annual rate of 14.5 per cent. In the early 1980s the international nitrogen market suffered from excess capacity and demand grew more slowly. Prices increased with demand from 1983 until mid-1985, when new sources of supply became available. In 1986 and 1987 prices remained low but in 1988 there was an increase to about $130 per tonne (f.o.b.). Studies by the FAO and UNIDO indicate that fertilizer prices have been volatile throughout the 1980s and may start to increase in the 1990s.[3] This implies that the netback to gas in the production of fertilizer in most instances could increase as prices increase.

In many developing countries, however, the building of new production capacity for fertilizer for domestic use is still an economically viable alternative to imports. It is difficult to generalize because the costs of fertilizer plants can vary by a factor of 25 per cent depending on the availability of local infrastructure and site preparation costs.

Estimates of the netback to gas for fertilizer production based on a typical world-scale plant (1,750 tons urea per day) show considerable variation. The main determinants of the netback are location (in terms of cost level and available infrastructure), the degree of export or import orientation, and fertilizer price projections. Netback figures vary from near zero for export-oriented plants in remote locations to as much as $3/mmbtu for favourably located plants with modern design and energy-saving features, aimed at local markets for import substitution.

(*c*) *Gas as a Methanol Feedstock*. Methanol, or methyl alcohol, is among the major basic chemical raw materials produced today. It is manufactured through a well-established chemical process based mainly on methane from natural gas. A limited part of world capacity is based on naphtha. There is a new interest in methanol due to a growth in its traditional markets through the recent development of new chemical applications. In addition, it has emerged as a possible means of reducing dependence on petroleum products in the transport sector.

[3] *FAO Fertilizer Handbook*, 1988.

The basic technology for producing methanol from natural gas is well established and has been improved to the point where no major breakthroughs in process efficiency are expected. Typical plant sizes are from 1,000 to 2,000 tons per day. Methanol can also be produced on floating plants. This may enable some developing countries to exploit offshore gas resources that have no alternative uses.

Applications of methanol are varied. More than 95 per cent of today's world methanol market is for chemical applications. These include traditional uses of methanol as a solvent and as a raw material for other chemicals such as acetic acid and single-cell protein. The most common use of methanol is in the production of formaldehyde, which accounts for about one-half of the total world market. Methanol can also be used as a fuel in a gasoline/methanol blend (up to 20 per cent methanol) in spark ignition engines, and for a variety of other purposes as fuel for power generation and as a household fuel. New chemical applications for methanol are being developed, including processes to convert methanol to ethylene, which is the most basic building block in petrochemical uses. If a breakthrough occurs, such a process will open up an enormous new market.

The netback to gas for methanol production is expected to remain low, owing to the persistence of excess capacity into the 1990s as well as the high cost of new plant. Total world capacity in 1982 was about 14 million tons and this is expected to double by the early 1990s once currently-planned projects are completed. World prices are therefore unlikely to offer a return sufficient to justify the construction of new plants except in countries with very low gas costs, a large domestic market for methanol and a suitable infrastructure and location for a new plant.

(d) *Petrochemicals and Polyolefins.* Unlike the production of nitrogenous fertilizer or methanol, petrochemical production is based on the use of the ethane, propane and butane portions of natural gas. It is therefore necessary to process the raw gas to remove these other components. Since the composition of natural gas from different fields varies widely, the determination of the minimum ethane content required to justify extraction in economic terms is an important consideration in petrochemical plants.

The most important stage in the use of gas in the petrochemical sector is the initial one of processing ethane or an ethane/propane mix in the steam cracker to produce (jointly) ethylene and propylene. The most important bulk intermediate products are low-density polyethylene (and linear low-density polyethylene), high-density polyethylene and polypropylene.

In their wide variety of applications, these polymers substitute for metals; timber, cork, paper and other wood derivatives; natural rubber; glass; and leather and fabrics. Polymers have also found new applications in agriculture and water management (storage ponds and canal lining for preventing water seepage, irrigation piping systems, seedling nursery bags, etc.).

One of the major advantages of using polymers is their contribution to energy efficiency, since the energy content of plastic products at the final application level is often significantly lower than that of competing products. This advantage is compounded where plastic products are recycled rather than abandoned after use.

The calculation of netbacks for gas in petrochemical uses is very complex and it is not possible to cite a general formula for it. It depends on the specific gas composition, the specific products being manufactured, and the local market conditions for these products. There are, however, many opportunities to expand gas use in the petrochemical sector in developing countries, since demand for many plastics and other products is expected to grow rapidly. Netbacks will be higher in countries that invest in world-scale plants and produce for the domestic market.

9 The Value of Gas in the Residential and Commercial Market

Many developing countries are experiencing a high rate of urbanization. The residential and commercial sector is responsible for a large and growing share of total energy consumption, primarily for cooking, but also for water and space heating. Countries that have gas distribution systems serving the residential and commercial sector are experiencing rapid demand growth, often surpassing the capacities of their gas distribution networks. For example, Pakistan has over 700,000 households connected to its natural gas distribution network. These consumers often experience gas shortages during peak demand hours, and many of the new applications submitted each year by households cannot be taken up because of the shortage of gas. Similarly, in Iran, when gas was first distributed in the early 1970s, demand increased much more quickly in the short run than the ability of the gas distribution company to meet it.

In the absence of gas, urban populations in developing countries rely on a variety of other fuels such as LPG, kerosene, gas oil, coal and fuelwood. The liquid fuels are generally premium products with a high cost; fuelwood supplies are diminishing and often unavailable in urban areas; coal can have high environmental and inland distribution costs. Natural gas is a clean-burning fuel which does not contribute to pollution. These conditions suggest there may be significant scope for expanding residential and commercial use of gas in developing countries.

9.1 Existing Residential and Commercial Gas Use in Developed and Developing Countries

Residential gas demand was about 26 per cent of total world gas demand in 1986. The largest residential market is in

Western Europe, where 40.5 per cent of natural gas use is in the residential and commercial sector, followed by the United States, where this share is 40 per cent.[1] Gas is available to 97 per cent of householders in the Netherlands and 75 per cent of householders in the UK.

The residential and commercial sectors in North America, Western Europe and Japan use large volumes of natural gas. In 1986, in the United States this sector was responsible for 40 per cent of total gas demand. In the Netherlands, which has large indigenous gas reserves, the share of gas used by this sector was about 45 per cent. Developing countries lag far behind OECD countries in the proportions of gas used by their residential and commercial sectors. In Pakistan, for example, this sector accounts for only 12 per cent of total gas consumption, while in Mexico and India it consumes less than 2 per cent. While demand for space heating, in particular, will be less in many developing countries, the city-gate cost of gas is also likely to be lower in many developing countries, which suggests that lower-value uses may be economic.

The historical record of residential and commercial gas use in the industrialized countries shows a pattern of rapid growth of demand following the initial infrastructural investment. Urban gas distribution took off in the late 1940s and was fully developed by the 1960s. For example, in 1975 in West Germany total consumption was 7.24 toe; by 1986 it was about 18 toe.[2] In some cities, natural gas followed 'town gas' which was manufactured from coal or naphtha. During the 1940s and early 1950s, most of the residential and commercial demand was for cooking, water heating, space heating and gas-fired refrigerators. Commercial use in schools, hospitals, hotels, restaurants, shops and offices also expanded rapidly. In the late 1950s and 1960s, gas demand grew as central heating became more widespread.

Developing countries with urban gas distribution systems include Algeria, Argentina, Bangladesh, Brazil, Egypt, India, Iran, Mexico, Pakistan, Tunisia and Venezuela. Several of these countries are planning expansions of their gas networks,

---

[1] *Le Gaz naturel dans le monde en 1988*, Cedigaz, 1989.

[2] Cedigaz, ibid.

The Value of Gas in the Residential and Commercial Market 103

and South Korea and Turkey may soon be added to the list. However, the existing distribution networks are all limited to certain suburbs of major cities and many of them are not linked to a country-wide grid. As in the OECD countries, once the infrastructure was put in place in developing countries, residential and commercial gas demand grew rapidly. Pakistan is one of the few developing countries with a relatively mature distribution system and accessible historical data. Since it started operating in 1955, the gas transmission and distribution system has been expanded to supply a current customer base of around 900,000 households and commercial users. The share of natural gas used by this sector increased from 4 per cent of total gas consumption in 1972 to about 15 per cent in 1987, and consumption grew at average annual rates, for the residential and commercial sectors taken separately, of 27 and 17 per cent respectively. In spite of this impressive growth, less than 25 per cent of the population in Karachi, Lahore and Islamabad has access to gas.

9.2 Estimating the Benefits from Residential and Commercial Use of Gas

It is conceptually straightforward to estimate the netback value of gas used in the residential and commercial sector. The results will vary widely, however, depending on the technical parameters and usage patterns of each city. Rather than using data from existing residential and commercial systems, which would not be readily comparable, we carried out a simulation study to explore the relationship between the technical and economic parameters of this use of gas in a city and the netback value of the gas.[3] In this way, system costs and economic netbacks for this use of gas can be varied to show how netback values change under a variety of circumstances.

Sixteen model distribution networks were developed to encompass the range of relevant combinations of parameters. To develop these model networks, gas demand was projected over a twenty-year period, and capital and operating costs were

---

[3] For a complete description of the study and its results, see Mashayekhi, A. and Sofregaz, Consultants, *The Economic Value of Natural Gas in Residential and Commercial Markets*, Energy Department Paper No. 22, World Bank, March 1985.

determined on the basis of the physical design parameters. Differences in urban characteristics, climatic conditions and consumption habits resulted in a large number of possible network configurations. The least-cost design was selected for each set of assumed parameter values. Table 9.1 shows the set of parameters used. They include two city types (existing and new), two levels of gas demand (base-load and base-load plus space heating), and four levels of population density (very high, high, medium and low).

(a) *Methodology and Parameters.* The least-cost simulations for distribution system design are based on a standard model area of 50,000 housing units. These are assumed to be located either in new suburbs or an existing part of a city. The optimal network design is somewhat shorter for the new suburbs because the lines can be laid in new areas instead of strictly following existing streets. It is assumed that the total length of the distribution system will be 20 per cent less in a new suburb

Table 9.1: Projected Levels of Annual Average Household Consumption for Various Model Distribution Networks. Volumes in Thousand Cubic Feet per Year per Household.

City Type	Gas Uses	Density	Consumption
Existing	Base-load (Cooking and Water Heating)	Very High	12.4
		High	12.4
		Medium	17.7
		Low	24.7
	Base-load and Space Heating	Very High	40.6
		High	40.6
		Medium	56.5
		Low	64.9
New	Base-load	Very High	12.4
		High	12.4
		Medium	17.7
		Low	24.7
	Base-load and Space Heating	Very High	40.6
		High	40.6
		Medium	56.5
		Low	64.9

than in an existing area. Costs are also reduced by co-ordinating the laying of pipelines with the installation of water and power lines, and by avoiding the need to dig up and re-lay roads. In addition, there are some savings in connecting the consumer since the purchase and installation of domestic appliances can be co-ordinated with the construction of the original building.

The second important technical parameter is total gas consumption. This depends on the number of residential and commercial customers, and the unit consumption of each. The latter is a function of price, income levels and the climatic conditions of the city. For the purposes of this study, the price is assumed to be constant over all cases, and climatic variations are represented by two alternatives: with and without space heating demand. Income levels are related to the assumptions made about housing density, as described below.

The third important technical parameter is the population density. Four cases were examined: very high density with 12,000 households/km^2 (e.g. Mexico City, Rio de Janeiro and Dhaka); high density with 6,000 households/km^2 (e.g. Tunis and Lahore); medium density with 3,000 households/km^2 (e.g. La Paz, Lagos and Istanbul); and low density with 1,500 households/km^2 (e.g. Warri in Nigeria, Santa Cruz in Bolivia and North Tehran). The number of domestic customers in all cases reaches 50,000 by the time the network is complete.

An important assumption for each of the density cases is that different consumption habits are displayed by collective and single-family housing units. Experience in developing countries suggests the existence of a high degree of correlation between the ratio of single-family to collective housing and the average income level. The latter, of course, is positively related to energy consumption. On the basis of observations in various cities, without space heating, households in the very high and high-density cases would each consume 12.4 mcf per year. This is equivalent to about two bottles of LPG per month. Households in medium-density areas were assumed to consume 17.7 mcf per year, those in low-density areas 24.7 mcf per year. By comparison, average annual household gas consumption is 62 mcf in the UK and 41 mcf in France and Italy.

The space heating category adds heating or air-conditioning demands to the above figures. This type of consumption has marked seasonal peaks, and this necessitates certain changes in the technical specifications of the distribution system. Experience from several countries indicates that gas requirements for space heating are about four times those for cooking and water heating. The figures used in this study are based on actual observations of 40.6 mcf per year for apartments and 64.9 mcf per year for single-family houses.

Consumption by the customers who are already connected is assumed to increase at an average annual rate of 1 per cent from the second year of consumption onwards. This rate of increase reflects actual operating experience in several countries. In all cases, the network's construction takes nine years to complete and customer connection starts after the first year of the project and lasts for about ten years following an S-shaped curve (see Figure 9.1). The rate of penetration, defined as the percentage of connections made by a given time, is an important determinant of the value of gas. If the construction schedule for the network proceeds as planned, but the speed of final consumer connections is slower, then actual consumption will fall short of planned consumption. Therefore, the actual penetration ratio has a major influence on the economics of gas distribution networks.

The distribution network costs include both non-customer-

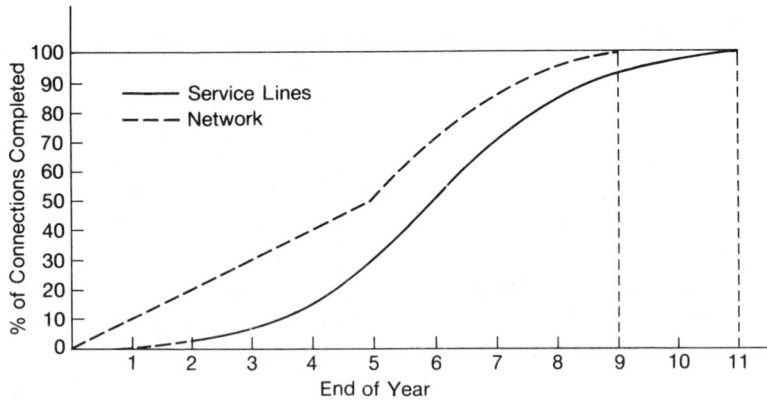

Figure 9.1: Construction Schedule for Network and Service Lines.

related and customer-related costs. Non-customer-related costs consist of the city-gate station, pressure regulating and metering stations, the steel pipeline network, pipe laying, fitting and design and supervision costs. These vary with the network's configuration, whose extent and length are functions of housing density. For example, the network costs for a low-density zone are about five times higher than those for a high-density zone. They also depend on the level of demand, particularly the demand for space heating. Space heating demand raises costs by 30 per cent but the gas flow is about three-to-four times greater than the base-load flow, thereby reducing unit costs overall by a substantial margin.[4]

Customer-related costs depend on the number of customers and population density, as well as the type of city. These costs include the external and internal services and depend on the type of housing, and the type and potential volume of consumption. The external services include all items outside the house or apartment; e.g. service lines, mains, customer regulators, meters, laterals, risers, etc. These costs are customer related in an existing city and non-customer related in a new city, where the decision to equip housing units for gas is made before the housing units are constructed. The internal service costs include the costs of converting appliances and connecting them to the gas supply in existing cities and the costs of new appliances in new cities. In single-family houses these costs are about 25 per cent higher than in apartment buildings. Appliance costs (based on observed costs in France) are taken as $91 for a cooker, $76 for a water heater and $270 for a radiator. These may overestimate the costs in some developing countries.

The commercial sector includes both large users such as hospitals, hotels, colleges and office buildings; and small users such as shops, bakeries, restaurants and laundries. Large users are assumed to represent 20 per cent of total commercial customers. The service line costs for the large users are similar to those of an apartment building. In the case of space heating the costs of a special regulating station are included. The internal services and appliance costs differ substantially for

---

[4] Mashayekhi, A. and Sofregaz, Consultants, ibid.

various commercial users. For simplicity, it was assumed that these costs for commercial users are similar to those for an apartment building. Costs for small commercial users were assumed to approximate those of single-family homes.

Operating costs include non-customer-related and customer-related categories. Non-customer-related operating costs cover the main network personnel and maintenance. Customer-related costs cover meter reading, bookkeeping, marketing, and legal work. These costs are all based on gas industry experience in developing countries.

The natural gas used by the residential and commercial sector displaces other fuels such as LPG, kerosene, gas oil, manufactured town gas, coal and non-commercial sources of energy. In many large cities, the main fuels displaced are LPG and kerosene. Observations in cities such as Cairo and Tunis indicate that LPG is the main fuel used for cooking and water heating. The netbacks are estimated using a discount rate of 10 per cent and assuming that the fuels displaced by natural gas were 80 per cent LPG and 20 per cent kerosene. Two price scenarios were used in estimating netback values. The base scenario assumes LPG prices of $200/ton and kerosene prices of $160/ton with both prices increasing at an average annual rate of 2 per cent after 1990. Additional handling and distribution costs of $150/ton for LPG and $30/ton for kerosene were included in the cost of alternative fuels since the cost comparisons are at the burner-tip. The high price scenario assumes an LPG price of $240/ton and kerosene prices of $300/ton.

There are no significant differences between the fuel efficiency ratios of natural gas and either LPG or kerosene.[5] The costs of marketing LPG in most developing countries are high, while the costs of marketing kerosene are much lower. All marketing costs are assumed to be constant in real terms throughout the period.

(b) Results and Sensitivity Tests. Table 9.2 summarizes the average incremental costs of distribution of gas to the residential and commercial sector. In an existing city with space heating,

---

[5] Construction is assumed to begin in 1983 and take place over five years. Operation begins in 1984 at a very low level and begins to level off by 1990.

Table 9.2: Average Incremental Costs of Gas Distribution. 1982 Dollars per Thousand Cubic Feet.

City Type	Density	Without Space Heating	With Space Heating
Existing	Very High	9.1	3.9
	High	12.5	5.1
	Medium	12.1	4.9
	Low	11.5	5.1
New	Very High	7.1	3.1
	High	8.5	3.6
	Medium	7.7	3.0
	Low	7.1	3.2

at a real discount rate of 10 per cent, these costs range from $3.90 to $5.10 per mcf; but these costs fall to a range of $3.00–3.60 per mcf in a new city. In the cases without space heating, costs are far higher: in an existing city as high as $12.50/mcf and in a new city as high as $8.50/mcf. The costs of exploration and production of gas and its transmission to the city gate must be added to these distribution costs to derive the total gas supply costs. In Chapter 5 we saw that the cost of gas supply to the city gate is around $1.50/mcf in many developing countries. This increases total costs of gas supply for residential and commercial consumers to a range of $4.61–14.00 per mcf.

It is worth noting that the distribution costs estimated for Bangladesh in Chapter 11 below are far below these costs. For example, distribution costs in the Bakhrabad system, which has no space heating, are about $1.20/mcf excluding connection costs. Even if connection, metering and other internal equipment costs are included, the Bangladeshi figures will still remain below the costs suggested above. Most of this difference can be attributed to low labour costs and different design factors which result in a lower standard of security of supply in Bangladesh. Distribution costs observed in many Latin American cities are towards the higher end of the spectrum. This indicates that the costs and benefits of gas use in residential and commercial markets must be assessed within a specific context, and provides no support for generalizations about the high cost of distribution infrastructure. In some instances costs are far below expected levels and benefits may be higher because of the higher handling and distribution costs of LPG.

The netback values of residential and commercial gas (excluding exploration, production and transmission costs) are presented in Table 9.3. There is a wide range of netback values for gas used in the residential and commercial sector. The netback values assuming the base price scenario range from less than zero to $3.37 per mcf. The negative values are found for supplying gas to less than very high density areas of an existing city.

Many developing countries such as Bolivia, Pakistan, Peru and Turkey, require space heating in some cities over a long winter period. Other countries such as Algeria, Morocco, Tunisia and some Latin American countries need space heating only during limited periods in certain areas. Many developing countries can also use gas for air-conditioning, as is done in Bahrain. Air-conditioning demand was not included in this study, although it would improve the economics of gas distribution projects where electric air-conditioning was the alternative.

Population density also influences the netback and net present values through its effect on total gas consumption. This effect is composed of two offsetting factors. High population density provides economies in the network construction cost

Table 9.3: Netback Values at the City Gate of Gas Used in the Residential and Commercial Sector. 1982 Dollars per Thousand Cubic Feet.

City Type	Density	Without Space Heating	With Space Heating
Existing	Very High	Negative	2.53
	High	Negative	1.23
	Medium	Negative	1.64
	Low	Negative	Negative
New	Very High	1.17	3.29
	High	0.00	2.70
	Medium	0.50	3.37
	Low	0.60	3.10

as well as the external services, but average gas consumption per household is generally lower in high-density areas. The higher consumption of single-family, high-income houses (in low-density areas) usually outweighs the effect of the lower costs of distribution to customers living in apartments.

Another important determinant of the netback value is the type of city. The netback is far greater in new areas of cities than in existing areas because distribution lines, external services, and internal services are provided at the construction stage.

The sensitivities of net present values and netbacks to whether a country is an oil exporter or an oil importer were also tested. The netbacks in oil-importing countries are higher by between 8 and 25 per cent than those in oil-exporting countries because of the difference between the c.i.f. and f.o.b. prices of the fuels that can be replaced by natural gas. Also, the present value of net benefits and netback values are very sensitive to discount rates, and fall quite drastically at higher rates. For example, in an existing city with a very high population density, and only base-load use, an increase in the discount rate from 8 to 12 per cent reduces the netback value by 28 per cent. Also, at a 10 per cent discount rate, there are some negative netback values in existing cities with high, medium and low densities and only base-load use.

9.3 Summary

The results reported in this chapter demonstrate that the use of gas in the residential and commercial sector in developing countries can provide significant economic benefits. The best case for gas use in the residential and commercial market is in new areas of cities with space heating demand. Density parameters affect netbacks, but overall impacts vary depending on the actual case. However, even for countries using gas only for base-load purposes, there are often significant economic benefits to be obtained provided it is technically possible to build a gas distribution network at reasonable cost.

PART IV

NATURAL GAS PRICING AND PLANNING

10 The Principles of Gas Pricing

Gas pricing can be a complex and controversial subject. For the most part, its complexities stem from the array of different considerations that come into play at the various stages where the gas changes ownership; from setting producer prices high enough to tempt the risk-takers into exploration and production, to creaming off some of the resource rent for the government, to finding equitable transfer prices between gas producers and utilities, to devising complex tariff structures to reward large consumers for shifting off peak or accepting a lower-quality service, to protecting small consumers from monopoly exploitation by the utility. Because the buyers and sellers are different at each level, different contractual arrangements will be appropriate. In the end, however, the whole chain of prices, from the gas explorer to the small consumer, must reflect, and properly distribute, the economic gains from gas development.

The controversies over gas pricing generally arise for two reasons. First, at most of the levels where gas changes ownership, either a monopoly or a monopsony prevails. There may be many producers in an area, but there is often a single utility buyer. There may be many potential consumers, but they have only one supplier. Thus, the unencumbered marketplace rarely provides clear signals to both parties on the value of the gas. It is for this reason that the gas market is generally regulated by governments. The second factor provoking controversy is the chain of transfer points that separate the first producer from the final purchaser. Once gas has been discovered, potential buyers and potential sellers see the prospect of capturing substantial resource rents, but only at the expense of each other. Left to itself, the inherent short-term and self-centred focus of each pair of buyers and sellers can be counter-productive to the success of the project as a whole.

Because of this potential for complexity and controversy, it is easy to lose sight of the basic economic principles that bear upon questions of gas pricing. The gas market contains many imperfections, and these violate the assumptions necessary for the derivation of some of the simplest principles of economic pricing. None the less, we believe that the basic theory is a useful starting-point, both to provide a common frame of reference into which modifications can be introduced and to confront the beliefs of those who contend that modifications (or market interventions) are unnecessary because the gas market can be expected to behave efficiently if left to its own devices.

In this chapter we discuss the objectives of gas pricing, the level of gas prices and the structure of producer and consumer prices.[1] The next chapter presents a detailed case-study of gas tariffication in Bangladesh.

10.1 Objectives of Gas Prices

Prices set by a central authority are generally based, implicitly or explicitly, on three sets of objectives:

(a) The efficiency of resource allocation;
(b) The satisfaction of specific financial targets;
(c) Considerations of social equity.

Sometimes conflicting objectives lead to contradictory results; for example, economic efficiency may imply the need to charge high rates to low-volume users while social equity could be furthered by charging subsidized rates to small consumers. Because such conflicts are common, it is important at the outset to review the relative suitability of gas tariffication as a tool for achieving each of the three objectives.

(a) *Economic Efficiency.* Economic efficiency requires that the real cost of energy to society be reflected in the prices that

---

[1] For a more detailed treatment of many of these see Munasinghe, M. and Warford, J. J., *Electricity Pricing: Theory and Case Studies*, The Johns Hopkins University Press for the World Bank, Baltimore and London, 1982.

guide the decisions of producers and consumers of that energy. When this principle is violated, energy consumption can rapidly get out of hand, particularly that of industrial and other large consumers. For them, in any case, the continuity and quality of the service are more important than its cost, which generally accounts for less than 10 per cent of their total budgets. The best alternative is to work at minimizing the cost of supply and promoting cheaper energy forms, such as off-peak gas, for their use. Subsidies that benefit large users translate into a serious financial handicap for the utility which, in turn, often leads to a deterioration of service which, in the end, is prejudicial to all.

(b) *Financial Viability*. A financially viable gas utility is one that has the capacity to finance, through its own resources and its ability to service borrowing, the expansion of its operations in line with the development of the market. Because the satisfaction of this criterion will depend on debts incurred in the past as well as the cost of future investments, the level of revenues needed to ensure the utility's financial viability will not necessarily coincide with that produced by the tariffs that would be economically efficient taking account of future costs alone. However, it is usually possible to devise a set of tariffs whose *structure* satisfies efficiency criteria and whose average *level* meets financial needs.

(c) *Social Equity*. The manipulation of gas tariffs to redistribute income often misses its mark. The tariff setter cannot substitute for the legislator: he has neither the political mandate nor the necessary information on consumers' income and assets. Furthermore, the correlation between the latter elements and gas consumption is usually unknown and sometimes perverse. For example, subsidizing the unit cost of gas encourages waste but fails to aid the small consumer who has few appliances, let alone the majority of poor households who have no access to gas. Such subsidies should be a last recourse to correct the deficiencies of direct income transfer mechanisms. They must be rigorously limited in order to safeguard the objective of financial viability, and carefully designed to satisfy, as closely as possible, the objectives of economic efficiency and social equity. For example, selective rebates on the connection charge, making access to gas easier, make up a subsidy that is more visible and more efficient than a reduction in the unit price of gas.

10.2 The Level of Gas Prices

Having said that the overriding criterion for gas pricing should be economic efficiency, the first step in setting or changing gas prices is clearly to determine the economic price of gas in the particular country or region in question. As described in Chapter 2, this will require a forecast of gas demand and supply over a medium- to long-term period. Depending on the relationship between demand and supply, it may also require an estimation of the depletion premium. This will be necessary in the 'surplus-window' case, as illustrated in the annex to Chapter 2. In the other cases – gas-short or gas-surplus – the economic price of gas can be derived more directly from its replacement value in the market-place (for gas-short regions) or its long-run marginal cost of production (for gas-surplus regions).

The economic price of gas is an essential bench-mark for pricing analysis, but it is only a starting-point. Gas prices must be viewed as a coherent *set* of signals linking all actors in the chain: producers, pipeline operators, distributors and consumers. Gas prices at one level should not be recommended without an analysis of appropriate corresponding prices elsewhere in the network.

In developing such a set of prices, there are three basic questions that must be confronted. The first is whether the framework should be set from the bottom up or from the top down. Should one start with the cost of production and add on margins, or pitch consumer prices correctly and work backwards? In the 1970s most industrialized countries used the former approach. Yet there are strong economic arguments for the latter approach, especially in gas-short countries. During the 1980s deregulation in the United States gas market and in some West European countries has led to top-down pricing. As a general rule, the economic price of gas is estimated at the point of consumption and using it as a bench-mark will therefore naturally lead to a top-down approach to pricing.

A closely related question is where the government should extract taxes – all through the line (from producers, distributors and consumers), only from producers, or only from consumers? This question is especially important for gas-short countries where the 'rental' element to be taxed away is large, but even in gas-surplus countries the government may wish to use the gas sector to raise revenue for the rest of the economy. Where should the tax wedge go? This question cannot be answered without taking into consideration the structure of other taxes in the economy, both on competing energy products and on producers (e.g. corporate profits tax) and consumers (e.g. income tax) more generally. However, as a general rule, the tax element should pushed as far up stream as possible; i.e. to the gas producer. This best preserves the signal of the economic price to the sets of buyers and sellers along the chain. The rationale for taxation through producer prices is discussed further in the next section.

The final question in developing an integrated set of gas prices is how to take account of other domestic energy prices in determining the appropriate 'pitch' of gas prices. In countries where the prices of competing fuels do not reflect their economic scarcity (because of either subsidies or taxes), it will be important to link the various levels of gas prices not only to each other but also to the competition. For example, if fuel oil prices are heavily subsidized but the government wishes to encourage gas exploration and development, it may be possible to finance consumer price subsidies by a profit tax on producers, while the producer price is set near the (higher) economic value of gas.

All of the above considerations relate to the determination of the appropriate gas price *levels* and the linkages among them. At each level, however, there are also important questions of the price structure and the related contract provisions such as take-or-pay, penalties for non-delivery, etc. With long-term investments being made by both producers and consumers of gas, there will be pressures from both sides to enter into guarantees on deliverability and purchases. Such guarantees, properly formulated, can speed the penetration of gas into existing and new markets by reducing uncertainties. Unless they are carefully designed, however, they can also negate the incentive effects of pricing. Therefore gas pricing policy should include the consideration of non-price clauses involved in the purchase and sale of gas.

10.3 The Structure of Producer Prices and Related Contract Provisions

Contrary to the case with oil, a major constraint on gas exploration is the absence of a predictable basis for forecasting the value to the producer of a possible gas discovery. Most exploration and production contracts do not include gas pricing provisions beyond saying that, in the event of a significant gas discovery, the investor and the government will negotiate a price. In this situation, there is no clear basis for the investor to evaluate the expected value of a gas discovery. Experience has jaundiced companies who, in the past, have been able to negotiate gas prices following discoveries that were adequate to justify only the incremental investments and insufficient to justify the initial high-risk exploration.[2]

There is general agreement on what the ideal contract provisions should provide:

(a) A predictable gas pricing formula which, prior to exploration, would permit an evaluation of the future value of a discovery;
(b) A price formula linking gas to its economic value in the market, thereby encouraging exploration whenever the expected costs are lower than the expected value;
(c) An efficient rent tax to retain the resource rents for the country.

Moreover, contract models from the oil industry can fairly readily be used to prepare clauses that satisfy these conditions

---

[2] This is generally not a problem with associated gas, where the investor will proceed with development to recover oil even if the gas price is very low. Contracts typically provide for government access to associated gas, free of cost, at the well-head if no commercial gas use is proposed.

for gas.[3] The thorniest part is generally the determination of the economic value of gas prior to knowing how much gas is likely to be found. However, that question is central to all work on gas pricing and planning, as discussed in Chapter 2. Various innovative contract provisions have been devised to accommodate the pre-exploration uncertainties, such as gas buy-back clauses, export guarantees in excess of the 'national reserve' and low base price and profit sharing for export-oriented industries.

10.4 The Structure of Consumer Prices and Related Contract Provisions

A detailed example of tariff design for gas consumers is presented in the next chapter. However, it is useful to consider the general principles for setting consumer prices before becoming immersed in the peculiarities of a specific case.

As with electricity, economies of scale in the gas industry mean that large, long-term investments will be made whose capacity will be only partly utilized in the early years. Under these circumstances there will often be large differences between short-run and long-run economic costs (see Chapter 4). Seasonality and peaking will be important economic concerns. Large consumers that are able and willing to accept interruptible service can significantly reduce system costs.

Such differentials in the economic cost of serving different consumers at different times will exist even if the general *level* of gas prices for a particular country is linked to a replacement fuel such as coal or fuel oil. Especially where the gas delivery system is still small and load diversity is low, signalling consumers to shift off peak can result in improved levels of reliability for all and, potentially, delay the need for new investment. It will also have significant financial benefits as it allows the infrastructure to be more intensively used in the early years.

---

[3] For examples see Palmer, K., 'Designing Contractual and Fiscal Frameworks for Gas Development', in Gault, J. C. (ed), *The Economics of Natural Gas in Developing Countries*, Pergamon Press, New York, 1985, or Stauffer, T. R. and Gault, J. C., 'Exploration Risks and Mineral Taxation: How Fiscal Regimes Affect Exploration Incentives', *The Energy Journal*, Vol. 6, Special Tax Issue, 1985.

The particular questions that should be addressed include:

(a) What are the demand characteristics (e.g. load factor, seasonality) of the major types of consumers? Given the cost of dual-fuel equipment to the consumer, for which should interruptible service be encouraged?
(b) What is the *structure* of the marginal cost of serving different types of gas consumer? How does it vary over time?
(c) What are the options for metering? Under what circumstances would demand, as well as consumption, metering make sense? Should block metering of small consumers be used?
(d) What kinds of incentives (price and otherwise) would be most effective in promoting rapid conversion to gas as it becomes available while minimizing the creation of undesirable precedents as the gas market matures?

These are mostly empirical questions whose answers will depend on the case under consideration. They are taken up in detail in the case-study in the next chapter.

In addition to the techniques of marginal costing, tariff setting requires an appreciation of the practical constraints affecting implementation. Simplicity of the rate schedule is important even where marginal costs are complex. The tariff-making authority must separate the essential differences from the less important, in order to develop a tariff that the consumers can readily understand and, therefore, take into account in making consumption decisions. The tariff setter should also keep in mind the constraints faced by the consumers when altering their consumption in response to differential tariffs. Thus, since it is easier for most consumers to alter their daily habits than their place of consumption, the differences in the rates should reflect the cost differences between peak and off-peak hours rather than the cost differences for consumers at different points along a given transmission line, for example. Rate increases should also be implemented more gradually for those consumers who are incapable of adapting in a short period of time.

The distortions that result from a simplification of the rate schedule should anticipate the response from consumers and the modifications that follow in the cost of supply. Rates with fixed brackets or distinctions by class of consumer often founder on this problem. Whether the rate is progressive, in order to shelter the small consumer, or regressive, in order to accelerate the recovery of capital, the crossing-over of a customer from one bracket or category to the next is usually dubious in terms of efficiency and fairness. In the light of technological advances in metering and billing, it may turn out to be easier to put more sophisticated rate formulae into practice.

Once the right compromise between simplicity and fairness has been found, the rate schedule will not precisely reflect all the complexities of costs and their variation over time. It will not perfectly inform the consumer of the value to society of raising or reducing his or her consumption. The utility may wish to complement the tariff schedule with public awareness campaigns and specific marketing actions.

The process of implementing new tariffs often proceeds in stages. First, it is useful to identify areas of data weaknesses and particular problem areas in the current investment plan and rate schedule. Financial projections should also be prepared at this stage, covering about a five-year period. Secondly, original projections are refined and a set of tariff options is drawn up. Thirdly, top management selects the preferred option and obtains approval from the relevant government bodies. Finally, generally in most cases the marketing department defines and carries out the transition policy towards the target rates.

This transition may be undertaken at different speeds for different types of consumer. While slowness is costly, it is also sometimes necessary in order to test consumer responses to the new formulae. It is generally better to push a revision slowly but steadily towards a target known to all, rather than to experiment and then turn back under the pressure of events.

The transition will sometimes have to take into account distortions in the prices of competing energy products or in supply facilities. Although the best response in the long run would be to eliminate these distortions, it may be temporarily necessary to sell gas below its economic cost if, for example, fuel oil is locally subsidized.

An argument frequently made to postpone the enactment of higher tariffs is that such price increases would spread throughout the economy and add to the rate of inflation. While it is true that gas is an important input for many industries, such an argument ignores the fundamental role that monetary policy plays in inducing or restraining inflation, as well as the effectiveness of competitive pressures – both from domestic substitutes and from internationally traded goods – in restraining cost-push pressures on prices. In addition, if the alternative to tariff increases is government subsidy of the gas utility, then the resulting increase in the government deficit is more likely than a tariff increase to provide an inflationary stimulus to the economy.

The underlying structure of the new tariffs will generally come from the 'phase-two' study of economic costs over the coming years. In Part II we discussed the rationale and methodology for undertaking such a study, and included the empirical results for eight countries. In the next chapter, the case of Bangladesh is used to demonstrate the steps from the marginal cost calculation to the design of gas tariffs.

11 Natural Gas Pricing in Bangladesh: a Case-Study

The purpose of this chapter is to illustrate the techniques described in Chapters 4 and 10 for marginal cost and tariff calculations. We have taken the Titas gas system serving the Dhaka area in Bangladesh as our example. After a brief introduction to the gas sector in Bangladesh, we estimate the marginal cost of gas in the Titas system. This encompasses exploration, production, transmission and distribution costs, broken down by major customer groups. Following the marginal cost estimation, we calculate the depletion premium and the set of economic prices of natural gas in Dhaka. These prices are then compared with the tariffs charged in 1986. The chapter concludes with recommendations about tariff design incorporating the structure of costs of gas supply and the fiscal objectives of the government.

11.1 Organization and Evolution of the Gas Industry in Bangladesh

The natural gas industry in Bangladesh started before independence with commercial discoveries of gas in 1955 and 1959 in the eastern part of the country. A 12-mile pipeline was constructed to supply gas to a cement factory in 1960. Meanwhile Petrobangla was established as a public sector company to handle oil and gas development, including the exploration, production, transmission and distribution of gas.

Since the mid-1970s gas production has increased steadily and the distribution system has expanded. The government has encouraged the substitution of gas for other fuels through competitive pricing. During the period 1976–86 gas consumption in the Titas system grew at an average annual rate of over 14 per cent.

This growth in consumption is supported by large natural

gas reserves – at least relative to the size of the market in 1986. Gas reserves have been estimated at about 12 tcf and condensate reserves at 18 million barrels.[1] However, the country's large population and history of rapid growth in energy demand suggest that, even with these large reserves, plateau production levels will be reached by the early 1990s unless additional gas discoveries are made.

11.2 Demand Forecasts

Several gas demand forecasts have been prepared for Bangladesh.[2] A forecast of the incremental 'average day' demand for the Titas system is shown in Table 11.1. The incremental demand for each plant was estimated for the power sector, which is the largest consumer of gas. The fertilizer industry is expected to be the second-largest user of gas. Two additional plants are expected to be built over the period up to the year 2000. The demand characteristics for other industries were estimated on the basis of the growth rates and load characteristics for existing plants and the locations and energy demand requirements of new plants.

In addition to the *average* incremental demand, the regularity with which the user takes gas from the network must be considered in cost and tariff studies. This involves the concept of modulation, or load factor. The load factor is the ratio of average to maximum or peak demand over a given interval of time. It may be calculated for a single customer, a group of customers, or the whole system, and on either a daily or an annual basis.[3] The modulation is important because the size, and thus the cost, of a system are largely determined by the peak flows it must handle. This means that over the course of a year the same total volume of gas can be delivered at a lower cost, the higher is the load factor, i.e. the more regular is the

---

[1] Unpublished consultant study by DeGolyer and MacNaughton, 1983.

[2] This chapter uses an adjusted version of the forecasts shown in *Bangladesh: Second Natural Gas Development Project*, Bechtel National Inc., Dhaka, Bangladesh, 1984.

[3] For further details, see Turvey, R. and Anderson, D., *Electricity Economics: Essays and Case Studies*, Johns Hopkins University Press for the World Bank, Baltimore and London, 1977.

Table 11.1: Incremental Average Day Demand Forecast in Titas System. 1989–2000. Million Cubic Feet per Day.

	Power	Fertilizer	Industry	Commercial	Residential	Total
1989	141.7	96.5	46.5	10.0	27.3	322.0
1990	166.0	96.5	51.8	11.2	30.3	355.8
1991	197.3	96.5	54.7	11.7	32.3	497.8
1992	220.8	96.5	57.0	12.5	34.0	420.8
1993	226.7	105.5	59.5	13.1	35.1	439.9
1994	241.6	119.0	62.2	13.9	37.9	474.6
1995	284.8	137.2	65.0	14.6	40.0	541.6
1996	313.8	137.2	68.1	16.2	42.3	577.6
1997	358.2	137.2	71.3	16.3	44.7	627.7
1998	373.9	137.2	74.9	17.2	47.3	650.5
1999	387.3	137.2	70.6	18.2	49.9	671.2
2000	451.7	137.2	82.6	19.2	52.8	743.5

Sources: Petrobangla;
Bechtel National Inc., *Bangladesh: Second Natural Gas Development Project*.

offtake. Because the peak demands for different customers do not always coincide, a 'coincidence factor' is used to estimate the customers' contributions to the system peak.

The peak hours in the Titas system are from 9 a.m. to 1 p.m. and from 6 to 11 p.m. Shortages often occur during those hours, however, so current consumption underestimates the level of actual demand and the *actual* peak period is probably longer than the *natural* system peak.

11.3 Marginal Cost Estimation

In order to derive an estimate of the marginal cost per unit of gas, the incremental demand figures developed above must be related to the cost of finding, producing, transmitting and distributing the incremental gas supply (see Chapter 4). This section takes each of these cost components in turn.

(*a*) *Exploration Costs.* Twelve gas fields had been discovered in Bangladesh by 1984. By the late 1970s all the private oil companies except Shell had relinquished their concessions, and there has been very little recent exploration. For the foreign oil companies it was uneconomic to develop gas under the prevailing market and pricing conditions. For Petrobangla the priority has been on the development of gas fields already discovered.

There are plans both to increase exploration efforts by Petrobangla and also to provide incentives to private oil companies to undertake exploration. Petrobangla's plans for the exploration and appraisal of the known fields include:

(a) Proving new reserves in the known fields by drilling appraisal wells to assess the size of existing discoveries;
(b) Further exploration in both the eastern and western areas of the country;
(c) A regional oil and gas study to highlight the areas to be explored in the future.

Seismic results indicate that the known fields in the eastern part of the country are located on regional tracts on which other hydrocarbon-bearing structures may be present. These structures are expected to contain sizeable quantities of gas. The cost of finding more gas in these areas is expected to be small.

An exploration programme by Petrobangla in the Bakhrabad area to prove 1 tcf of gas reserves, consisting of profiling some 500 line-km of seismic surveys and drilling six exploration wells, would cost about $45 million. Assuming that this programme provides four producing wells out of the six drilled, and moves reserves of the order of 1 tcf from the probable to the proven reserve category, it would give an average exploration cost of $0.05/mmbtu.

Exploration in new areas will have a much lower probability of finding gas and may also be more costly. The limited experience of exploration in new areas makes the estimation of exploration costs highly uncertain. The western and southern parts of Bangladesh have been only very lightly explored, partly because adverse surface conditions make seismic work expensive. The chances of finding gas in these areas are good, but exploration costs are expected to be several times greater than the $0.05/mcf calculated above for the Bakhrabad area.

(b) *Production Costs.* At present the production of gas is mainly from the Titas, Habiganj, Bakhrabad and Sylhet fields. The Titas gas field, in operation since 1968, is the major supplier of natural gas to the Titas franchise area. During 1984 the field had six producing wells with a total capacity of 100 mmcf/d. With the completion of new wells, production has increased, and the condensates are separated from the gas at the wellhead at a rate of 1.4 bbl/mmcf.

The Habiganj field, also in operation since 1968, had two producing wells with a total capacity of 60 mmcf/d in 1984. Upon the completion of two new wells, production is about 100 mmcf/d with a condensate recovery of 0.05 bbl/mmcf. The Bakhrabad gas field started to produce in 1984. It has five producing wells with a total capacity of 150 mmcf/d and a condensate recovery of 2 bbl/mmcf. The Kailashtila gas and condensate field, in operation since 1983, currently produces up to 30 mmcf/d with a condensate recovery of 12 bbl/mmcf. There are no plans for new drilling in the Sylhet field.

There are major shortcomings in the Titas system. Existing production capacity is vulnerable to disruptions and there is no spare production capacity in the Dhaka area to prevent gas outages. Extra wells or gas storage facilities are needed to meet the demand for gas and to increase the system's reliability in the event of technical problems, unplanned repairs or emergency shut-downs. Spare wells are needed to ensure a reliable gas supply system and allow for necessary workovers, tests and maintenance. Spare gas processing capacity would also increase the processing reliability.

To satisfy demand in the Titas area, gas will be supplied by:

(a) Further development of the Titas and Habiganj fields by drilling two extra wells in each field to provide an additional 120 mmcf/d of gas;
(b) Linking the Bakhrabad and Titas systems to take advantage of the spare production capacity in the Bakhrabad system in the short run, and drilling additional wells in Bakhrabad to meet demand in the long run;
(c) Development of the three smaller gas fields.

The growth in demand in the Dhaka area is expected to be

Table 11.2: Marginal Cost of Expanding Natural Gas Production at Titas and Habiganj Fields.[a] 1984–2000. Values in Million 1986 Dollars.

	Incremental Costs			Incremental Production					Condensate Revenues[d]
	Capital	Operation and Maintenance	Total	Gas[b] Volume bcf		Condensate[c] Volume '000 bbl			
				bcf	bbtu	'000 bbl	bbtu		
1984	–	–	–	–	–	–	–	–
1985	14.0	–	14.0	–	–	–	–	–
1986	14.0	0.7	14.7	–	–	–	–	–
1987	–	1.4	1.4	19.8	20,550	27.6	140.8	0.4
1988	–	1.4	1.4	39.6	40,550	29.7	150.9	0.5
				39.6	40,550	29.7	150.9	0.5
1989	–	1.4	1.4	39.6	40,550	29.7	150.9	0.5
1990	–	1.4	1.4	37.6	40,550	29.7	150.9	0.5
1991	–	1.4	1.4	37.6	38,928	28.5	144.9	0.5
1992	–	1.4	1.4	36.1	31,371	27.4	139.1	0.6
1993	–	1.4	1.4	34.7	35,870	26.3	133.6	0.6
1994	–	1.4	1.4	33.2	34,441	25.2	128.2	0.6
1995	–	1.4	1.4	31.9	33,064	24.2	123.1	0.6
1996	–	1.4	1.4	30.6	31,741	23.2	118.1	0.6
1997	–	1.4	1.4	29.4	30,472	22.3	113.4	0.7
1998	–	1.4	1.4	28.2	29,252	21.4	109.8	0.7
1999	–	1.4	1.4	27.1	28,083	20.6	104.5	0.7
2000	–	1.4	1.4	26.0	26,959	19.7	100.4	0.7
NPV[e]	23.6	8.0	31.6	205.2	211,365	164.9	838.3	3.3

AIC (raw gas) = $0.16/mmbtu
AIC (lean gas) = $0.15/mmbtu
AIC (lean gas) for Habiganj only = $0.15/mmbtu

Notes: (a) Based on the drilling of two wells in Titas (60 mmcf/d), two wells in Habiganj (60 mmcf/d) with treatment facilities for the total capacity of 120 mmcf/d.
(b) The calorific values of the Titas and Habiganj gas fields are 1,038 and 1,010 btu/scf respectively.
(c) The calorific values of Titas and Habiganj condensates are estimated at 5.1 mmbtu/bbl; condensate production is 1.4 bbl/mmcf for Titas gas and 0.1 bbl/mmcf for Habiganj gas; 99 per cent of the cost per btu is allocated to gas.
(d) Condensate revenues are based on a condensate value of crude oil multiplied by a factor of 1.4 to estimated oil prices of $14.00/bbl. Crude oil prices are expected to remain constant in real 1986 dollar terms until 1990, and then to increase at an average annual rate of 2 per cent.
(e) Based on a discount rate of 12 per cent.

met by a combination of these three options. The cost of natural gas production differs depending on the source of gas as well as the methodology used to allocate joint costs between gas and condensates. The expansion of the Titas and Habiganj fields (Option (a)) will produce both gas and condensates. The main product of these fields is gas; condensate is a by-product requiring separation and treatment facilities as well as a separate transport system. As illustrated in Table 11.2, the cost of the raw gas produced is $0.16/mmbtu. Because the value of the gas is not known and the joint costs cannot be allocated according to value, the revenues from the condensate are deducted from the total gas costs. This results in lean gas costs for Titas and Habiganj of only $0.15/mmbtu.

Similarly with the drilling of new wells in Bakhrabad (Option (b)), natural gas will be the main product while condensates are the by-product. In this case, the incremental cost of raw gas from Bakhrabad is, as in Titas, $0.16/mmbtu. The cost of lean gas, if condensate revenues are subtracted from total production costs, is $0.11/mmbtu.

The development of the three smaller fields, Kailashtila, Beani Bazar and Rashidpur (Option (c)), would provide large revenues from the sale of condensates. The raw gas cost, as indicated in Table 11.3, is $0.16/mcf. However, if the condensate revenues are subtracted from total costs, the net lean gas costs are zero.

The share of the existing fields in total future production to the year 2000 will probably be quite large. The incremental gas to the Titas system is and will continue to be from the Titas and Habiganj fields (Option (a)). In the south-east part

Table 11.3: Marginal Costs of Gas Production.[a] 1986 Dollars per Thousand Cubic Feet.

Option	Raw Gas	Lean Gas
(a)	0.16	0.15
(b)	0.16	0.11
(c)	0.16	0.00

Note: (a) The marginal cost of raw gas includes all joint costs; the marginal costs of lean gas is obtained by subtracting the revenues from condensate sales.

of Bangladesh, the incremental supply of gas is expected to be from the Bakhrabad fields (Option (b)). In the north-east, the incremental supply will be from the three smaller fields. Once these three systems are interconnected, non-associated dry gas from the Habiganj field will provide the marginal source of natural gas to the entire system. Thus, for the purposes of this study, the net cost of Option (a) is taken to represent the incremental production cost of gas for the Titas system.

(c) Transmission Costs. The Titas transmission system cannot meet peak hourly demand. New additional transmission facilities are planned in the Titas area, and they can be virtually fully utilized to meet the existing demand with little further investment in distribution to connect new customers. The distances from the fields to the city are short, and there is no extra capacity in the pipeline system into which to 'pack' gas during off-peak hours and 'unpack' it to meet hourly demand fluctuations. In practice, the system operates at its peak capacity most of the time. Therefore capacity costs are estimated and allocated according to peak-hour use. The lowest load factors are for residential and commercial users.

Table 11.4 shows the investment and operating costs related to the extensions in the Titas transmission system. These costs are low because the only incremental costs are for

Table 11.4: Present Values of Average Incremental Cost of Transmission and of Demand for Titas System. Demand Costs in Million 1986 Dollars.

Present Value of *AIC*	
Investment	111.8
Operation and Maintenance	7.5
Total	128.3
Present Value of Demand (mcf/h)[a]	91.0
AIC of Transmission	
$/mcf/h	1,447.6
$/mcf	0.36

Note: (a) We use the peak hour (mcf/h) as opposed to peak day (mcf/d) as the measure of peak system demand.

extension of the existing system, and consumption is expected to increase since demand currently outstrips supply.

(d) Distribution Costs. Gas distribution costs are generally estimated for the network as a whole. It is sometimes possible to divide costs into capacity, or demand, costs, which depend on the diameter of the pipeline, and customer costs, which depend on the cost of providing consumers with access to the gas system. Estimates of customer costs are based on the minimum grid approach. Theoretically, the cost of area coverage is the cost of building a network with zero capacity or with sufficient capacity to serve only the requirements of appliances' pilot lights.[4] This method is often used by gas utilities. The advantage of this approach, compared with the alternative of lumping all the distribution costs together, is that it allows the separation of peak demand or capacity costs from the average connection costs for providing access to the service.

In practice, however, for any system the division between the customer- and the capacity-related costs is often done arbitrarily. The allocation of operating and maintenance costs is also difficult because, for many items, costs related to the addition of new customers to the distribution network cannot be separated from those related to the addition of new capacity needed to meet increased demand.

As an example of the minimum-size approach, the cost of minimum distribution network in the Titas area is provided in Table 11.5. It is based on the actual figures for Titas until 1988 and forecasts to the year 2000. The customer-related investments for years beyond 1988 were developed using historical average costs, and assuming that the initial investment in gas lines would serve all customers connected in the first four years until 1988 (as has proved to be the case) plus 80 per

---

[4] In the telecommunications sector, where expansion programmes increase both the number of connections and the number of calls, it is possible to separate the subscriber-related costs, including the cost of connecting to the system and being able to use the system once in each billing period during off-peak hours, from the traffic- (capacity-) related costs of making additional calls. See Saunders, R. J., Warford, J. J. and Wellenius, B., *Telecommunications and Economic Development*, Johns Hopkins University Press for the World Bank, Baltimore, 1983.

Table 11.5: Customer-related Distribution Cost in the Titas Franchise Area for Different Pipeline Sizes. Thousand 1986 Dollars.

Local Area	12–14 Inch	10 Inch	8 Inch	6 Inch	4 Inch	3 Inch	2 Inch	1 Inch	¾ Inch	Total
Jinjira	4	–	50	–	4	–	10	10	100	178
Savar–Manikanji–Aricha	179	–	–	–	10	–	15	25	30	239
Joydevpur–Tangani	–	220	–	–	–	30	40	50	100	470
Kaliganj	–	–	–	75	45	–	10	15	10	125
	183	220	50	75	59	30	75	100	240	1,032

cent of those connected up to 1996. It was also assumed that all customers connected after 1996, as well as 20 per cent of customers added between 1989 to 1995, would require extensions to the mains.

The total length of the distribution system described in Table 11.5 is 415,000 feet. The minimum size pipe is $\frac{3}{4}$ inch which costs 15 taka per foot to purchase and 22.1 taka per foot to lay, giving total cost of 37.1 taka per foot. Multiplying this by the length of the system gives a total cost for a minimum system grid of 15.4 million taka, or $616,000. This represents about 60 per cent of the actual total cost given in Table 11.5.

Apart from the minimum system grid, customer costs also depend on the number of customers served. The average cost of street mains (service mains) required per customer generally varies with the type of customer served. Domestic consumers require the shortest average length of connecting line because they generally have smaller properties and are concentrated in multiple-family dwellings; commercial service is more costly to construct. Industrial and other bulk users have the highest-cost service mains. When divided by the volume of bulk consumption, however, this cost is very low. For the Titas system, factors for commercial and industrial utilization of ten and twenty times the average cost of residential services, respectively, were used. Operating and maintenance costs were also allocated according to the relative weights of customer- and demand-related costs.

Aside from area coverage, the major design criterion for a distribution system is the peak demand. Ideally, the costs of distribution and feeder mains should be identified separately from the costs of street and service mains that provide the access function. The feeder mains are generally routed through the major industrial areas so that the larger requirements of industrial customers can be met from the larger and higher-pressure pipelines. Available data do not allow the separation of these figures for the Titas system Instead, the demand-related costs were estimated by subtracting incremental customer-related costs from total distribution costs.

The present values of incremental distribution costs up to the consumer's gate are presented in Table 11.6. These figures are divided by the present value of consumption or the num-

Table 11.6: Incremental Distribution Costs for Titas System. Present Values in Million 1986 Dollars.

	Demand-related	Customer-related	Total
Present Value			
Investment	7.5	0.5	8.0
Operation and Maintenance	0.2	0.5	0.7
Total	7.7	1.0	8.7
Average Incremental Cost			
$/mcf	0.2	0.03	0.23
$/customer	n.a.	2.18	2.18

ber of customers to obtain the figures in the bottom part of the table.

Both the demand- and the customer-related costs for the Titas system are far below those experienced in the developed countries, which have higher labour costs and stricter standards of security and reliability. A study of gas distribution costs by Sofregaz indicated that over 60 per cent of total distribution costs represent the cost of labour.[5]

In addition to these distribution costs, there are the connection costs for internal service lines from the mains to the house or the plant, meters, etc. All customers also pay for internal pipeline costs. For an average residential customer this is about 1,000 taka ($40). Such costs vary significantly from case to case for commercial and industrial customers. Customers are also charged separately for their actual connection costs. On average, these costs are $47 for residential customers, $60 for commercial customers, and $1,163 for industrial customers.

(e) *Cost Summary*. Total Titas system costs are provided in Tables 11.7 and 11.8. The average incremental cost for the Titas system as a whole is $1.56/mcf. Costs differ between consumer categories because of differences in transmission

---

[5] 'Study of Economic Costs and Benefits of Natural Gas Utilization in Residential and Commercial Markets', Sofregaz, 1983.

Table 11.7: Average Incremental Costs for Titas System. 1986 Dollars.

	Residential	Commercial	Small Industrial	Total
Present Value of Demand				
Capacity (mcf/h)	6,037	1,373	34,200	41,610
Volume (mmcf)	6,580	2,721	29,078	38,379
Customer Units	290,569	79,655	100,892	471,116
Average Incremental Costs ($/mcf/h)				
Transmission	1,042.3	1,042.3	1,042.3	1,042.3
Distribution	185.4	185.4	185.4	185.4
Sub-total	1,227.7	1,227.7	1,227.7	1,227.7
Cost per Customer Unit	2.18	2.18	2.18	2.18
Average Incremental Costs ($/mcf)				
Capacity Costs:				
Transmission	0.95	0.53	1.22	1.13
Distribution	0.17	0.09	0.22	0.20
Sub-total	1.12	0.62	1.24	1.33
Customer-related Cost	0.10	0.06	0.01	0.03
Sub-total	1.12	0.68	1.45	1.36
Exploration and Production Costs	0.20	0.20	0.20	0.20
Total	1.42	0.88	1.65	1.56

Table 11.8: Summary of Marginal Costs of Gas in the Titas System. 1986 Dollars per Thousand Cubic Feet.

	Residential	Commercial	Small Industrial	Total
Exploration	0.05	0.05	0.05	0.05
Production	0.15	0.15	0.15	0.15
Transmission	0.95	0.53	1.22	1.13
Distribution:				
Capacity-related	0.17	0.09	0.22	0.20
Customer-related	0.10	0.06	0.01	0.03
Total	1.42	0.88	1.65	1.56

and distribution costs resulting from differences in construction costs, population densities, composition of consumer groups and consumption levels.

11.4 The Depletion Premium

As explained in Chapter 2, the economic cost of a depletable resource such as natural gas includes two components: its marginal cost and a depletion premium. The latter is incurred whether the gas is consumed during peak hours or off peak. Current consumption of a depletable resource is at the expense of future consumption regardless of the time of day at which it takes place. That opportunity cost of forgone future consumption, i.e. the depletion premium, must be properly reflected in the level and structure of the tariff, in addition to the marginal costs calculated above.

In order to calculate depletion costs we used a simple iterative model – as illustrated in the annex to Chapter 2 – with sensitivity tests to allow different assumptions for: (a) the volume of reserves; (b) the gas production pattern; (c) alternative fuel prices; and (d) the discount rate. With reserves of 12 tcf, a production profile that starts to decline after twenty years, a discount rate of 12 per cent and a replacement fuel cost of $3.50/mcf, the depletion premium is $0.06/mcf. This must be added to the marginal cost of gas shown in Table 11.8 to yield the economic cost of gas in Dhaka.

140 *The Economics of Natural Gas*

The depletion premium is sensitive to the discount rate, the levels of demand and supply and the value of the replacement fuel, but is not very sensitive to the marginal cost of gas. If the price of the replacement fuel increases from $3.50 to $5.00 per mcf, then the depletion premium increases from $0.06 to $0.09 per mcf. If a lower reserves estimate of 5 tcf is assumed (or, alternatively, a much higher rate of growth of demand), then the depletion premium increases to $0.30/mcf. In the case of Bangladesh it is more likely that the reserves will turn out to be higher – rather than lower – than the 12 tcf assumed above. If future exploration reveals reserves to be 20 tcf, then the depletion premium will fall to $0.03/mcf.

11.5 Comparison of 1986 Tariffs with Economic Costs

In 1986 the tariff structure in Bangladesh consisted of a constant metered rate per unit of gas consumed for commercial and industrial users and a flat rate for residential consumers based on the number of cooking and water-heating appliances in the house or apartment. There was no seasonal or hourly peak rate. Consumers paid a lump-sum connection fee and there were no separate charges for metering or other customer-related costs.

As indicated in Table 11.9, the 1986 level of gas prices was remarkably close to the economic cost for residential users, almost 90 per cent higher than the economic cost for commercial users and 15 per cent below the economic cost for small industrial users. However, the sales price included a large excise duty imposed by the government in order to capture a part of the resource rent. Thus the price received by the utility for the gas it sold to residential and industrial users was far below the economic cost. The utility often had to apply to the government to cover its operating deficit, and made no contribution to its own investment needs. The problem of these low prices goes further up the chain to the gas production companies, which often incur losses despite the low cost of producing gas in Bangladesh and its high value in the marketplace. Despite excess demand, they are provided with no incentive either to explore for additional reserves or to undertake any new investment to expand production.

Table 11.9: Comparison of Existing Gas Tariffs with Economic Costs. June 1986. 1986 Dollars per Thousand Cubic Feet.

	Residential	Commercial	Small Industrial
Marginal Cost	1.42	0.88	1.65
Depletion Premium	0.06	0.06	0.06
Economic Cost	1.48	0.94	1.71
Sales Price	1.49	1.77	1.47
Excise Duty	0.45	0.61	0.55
Price to Utility	1.04	1.16	0.92
Alternative Fuel Prices[a]	5.00	5.00	3.20

Note: (a) Based on the 1986 c.i.f. value of kerosene in residential and commercial uses and the c.i.f. value of a 50 per cent fuel oil, 50 per cent gas oil combination for the small industrial users.

In order to provide appropriate signals to the consumer, the structure of gas tariffs must reflect the structure of economic costs. In the case of Titas, where capacity constraints are particularly important, this implies that a two-part tariff would be necessary to differentiate between peak and off-peak use. The energy component should include exploration and production costs and the depletion premium. It should be charged to all consumers in proportion to their use. The second part of the tariff should be a capacity cost to be charged to peak gas users. At present most consumers are using gas at peak hours, and the gas networks are operating at full capacity over a long period of the day. With the expansion of the system, however, the peak will reflect demand patterns rather than supply constraints. The inclusion of peak and off-peak rates would compensate customers with high load factors and encourage those with low load factors to make changes to their patterns of gas use where possible.

In order to introduce capacity charges, the larger power and industrial customers should be metered on an hourly basis or offered interruptible service at a lower rate. For the smaller commercial and residential consumers, hourly metering would be too expensive and administratively complex.

These consumers also have a known pattern of gas use for water heating and cooking which is difficult to manipulate. It would seem reasonable to continue with a flat charge that reflects energy costs (exploration, production and depletion), a capacity cost related to the number of burners and peak demand, and a lump-sum or annual fee for connection and internal service.

Gas tariffs must take account of financial and fiscal objectives as well as economic ones. Without significant additional information it is not possible to calculate the average price that the utility needs to receive for the gas it sells (i.e. after excise taxes) in order to cover its costs and make a reasonable contribution to its own investment programme. For the sake of this case-study, however, it is reasonable to assume that financial viability would be achieved if the utility received an average revenue equal to the marginal cost of gas, shown in Table 11.8 as $1.56/mcf. In an expanding system, marginal costs are often above average costs, so this is a generous assumption. Such a price should permit the utility to pay a marginal cost-related price to the gas producers.

It is also impossible from the information above to determine the appropriate level of taxes that the government should levy on the gas consumer to capture a portion of the resource rent. Certainly it should be no lower than the depletion premium of $0.06/mcf. However, given the large gap between the economic cost of gas and the prices of alternative fuels, the tax could be much higher. For simplicity we assume that the current levels of excise taxes properly reflect the government's fiscal objectives as they relate to natural gas.

Social considerations are often cited to justify low prices for gas. In Bangladesh, however, there is little support for this argument even though income levels are very low. Petroleum products are not subsidized and lower-income urban and rural groups without access to natural gas must use those more expensive fuels for their commercial energy needs. An expansion of gas distribution into poorer urban neighbourhoods, which would be possible if financed by higher gas prices, would be more beneficial to the poor than a general subsidy to existing (wealthier) consumers. On the basis of these considerations a possible gas price structure for the

Table 11.10: Possible Gas Tariff Structure for Titas System. 1986 Dollars per Thousand Cubic Feet.

	Residential	Commercial	Small Industrial
Energy Cost[a]	0.26	0.26	0.26
Capacity-related Cost[b]	1.12	0.62	1.44
Customer-related Cost[c]	0.10	0.06	0.01
Price to Utility	1.48	0.94	1.71
Excise Duty	0.45	0.61	0.55
Sales Price	1.93	1.55	2.26

Notes: (a) Includes exploration and production costs and depletion premium.
(b) Includes transmission and distribution costs.
(c) Excludes internal service and connection costs which are charged as a lump sum.

Titas gas system is shown in Table 11.10.

The upper limit to the sales price is set by the price of the alternative fuel adjusted for transport differentials. This limit is not reached by the proposed structure. For example, even during the peak period the small industrial user pays $2.26/mcf for gas while his alternative fuel cost is $3.20/mcf (see Table 11.9). For residential and commercial users the consumer surplus is even larger.

The impact of the proposed tariff reform varies by class of consumer. For residential consumers, the price would increase by 30 per cent unless the government chose to forgo the tax on this group. For commercial users the proposed tariff is actually 12 per cent lower than the existing price; the government might well choose to increase its tax take from this group and keep its sales price constant. Industrial users would see their unit tariff go up by about 55 per cent during the peak period, when all capacity charges are included, and down by about 45 per cent for off-peak use. If this differential had the desired effect of shifting some use from peak to off-peak periods, then the marginal cost would fall to reflect the investment deferred. These proposed tariff changes would yield the same revenue for the government while providing financial viability for the utility and an adequate return to gas exploration and production companies.

12 A Practical Approach to Gas Planning

One of the main themes of this book has been the importance of linkages between the various aspects of gas development. A market forecast is needed before the marginal cost of gas production can be calculated. That cost is generally an important factor in setting gas tariffs. The tariffs, in turn, will influence market demand. Projections of both demand and supply are needed to estimate the depletion premium for gas, which is also important in setting prices.

This situation has many parallels in the electric power sector. A hydroelectric site may be so large that the cheap power it offers cannot be economically developed until the domestic market for electricity grows much larger. Variations of demand during a typical day or across the seasons of the year determine the mix of supply investments needed to meet both peak and average levels of demand at the lowest cost. The choice of new plant will depend on the inherited structure of the power system as well as its expected future rate of growth. For example, a proposed hydro project will look more attractive if the inherited power system is mostly based on oil than it would if the inherited system were based on hydro and coal. In the latter case, the same hydro project would result in lower savings of fuel costs.

Because of these kinds of interaction among decision variables, investment planners must use an analytical framework that is multi-dimensional and which incorporates historical as well as expected variables. In the power sector, a large number of mathematical planning models with these features have been developed since the mid-1950s. It is now routine to use such a model of the power system in order to compare the total system costs of alternative new investments.

In the gas sector, by contrast, we are not aware of any formal investment planning models that incorporate demand

146 *The Economics of Natural Gas*

and depletion analysis.[1] Investment decisions tend to be made piecemeal and in response to local conditions of excess demand at the prevailing market price or newly discovered associated gas at near-zero incremental cost. Too often, in developing countries, only such associated gas is being used domestically. Once flaring has been eliminated, there is no further development of the national gas market even when there are substantial potential demands from the power and industrial sectors and large reserves of non-associated gas. Planning for such development requires different expertise and a more complex analytical framework. Without those tools, the supposed 'uncertainties' overwhelm even the sophisticated potential investor.

12.1 Questions to be Addressed

It is clear that embarking upon the development of natural gas – a non-renewable fuel and one that requires large, up-front investments for its transportation and use – raises complex questions of gas allocation and investment strategy which must be faced at the pre-investment stage of development. Many of these questions are similar across countries. For example:

(a) Should gas be used in electricity generation to replace imported fuel oil or coal, or should a gas-based fertilizer plant be built to replace imported urea?
(b) Would the high cost of a city gas distribution network be justified by the very high and growing cost of the kerosene and LPG used by households that could be replaced by gas?
(c) If gas reserves are large, should the country try to attract commercial partners for an LNG export project or would it be better to keep the gas in the ground to satisfy future growing domestic markets?

---

[1] The nearest attempt, although with a different focus, is the energy planning model as reported in deLucia, R. J., Jacoby, H. D. and others, *Energy Planning for Developing Countries: A Study of Bangladesh*, Johns Hopkins University Press, Baltimore and London, 1982.

A Practical Approach to Gas Planning   147

(d) As a producer of fertilizers or petrochemicals for export, could the country compete with supplies from the Middle East or Mexico where production is based on associated gas?

(e) Should gas storage facilities be developed for associated gas during the off-peak season in order to keep oil production constant, or should the excess off-peak gas be flared?

There are two things to be noted about such questions. First, they are essentially economic rather than technical questions. The technical feasibility of using gas for power generation, for fertilizer production and so forth has been well proven and need not be established anew. The important issue is that of the relative economic merits of the different alternatives. These will depend upon such country-specific parameters as the amount of base-load hydro in the power system, the proximity to major export and import markets for urea and the density of housing in urban areas.

The second point to be noted is that, although such questions concern specific alternative projects, they are really about long-run sector strategy. What should the role of gas be in a country with abundant lignite or hydro resources? Can gas be used as a engine of economic growth through export as urea or LNG? In which sectors can the penetration of gas as an oil substitute be most rapidly achieved? How could such substitution be phased out later if supplies become scarce and can be allocated only to the highest-value uses?

Because gas is depletable, such trade-offs over time, as well as the trade-offs between different gas-using projects at any point in time, must be explicitly considered. Further, because gas infrastructure is lumpy, it should generally be designed on the basis of total, long-run gas demand rather than that of individual gas-using projects. For these reasons, just as in the power sector, a long-run, sector-wide framework is generally necessary for the evaluation of individual projects and for the design of individual gas investments.

A second objective in developing such a sector-wide framework is to provide a way of testing the potential economic feasibility of gas-using and gas-producing projects at an early stage. Despite the large number of uncertainties typically facing the gas planner in the early days of development, it is important to undertake a rigorous pre-selection analysis at that time while the quality of estimates for different projects is comparable. The aim of this analysis should be to exclude the non-starters as much as to identify the clear choices (upon which full feasibility or preliminary design studies can begin). It should also define the critical, but currently uncertain, parameters that affect individual projects, so that these can be explored more thoroughly in the next round of feasibility or design studies.

12.2 The Scope and Focus of a Gas Planning Model

The appropriate scope and focus of a gas planning model (GPM) will depend on the complexity of the country's energy sector and the size of its gas reserves relative to domestic demand. For a country such as Thailand, with relatively small gas reserves and alternative indigenous resources of lignite and hydro, a fairly sophisticated model would be necessary. On the other hand, for Nigeria, with its huge gas resources and fewer indigenous alternatives, a simpler GPM would suffice. Indonesia, with large gas reserves but also major coal and hydro resources, would again require a more detailed model.

A GPM should not be expected to make final project selection decisions for the country's gas development programme over, say, a period of two decades. That would be impossible, given the large and numerous uncertainties inherent in pre-investment analysis. Rather, the GPM should be regarded as a preliminary, though rigorous and comprehensive, planning tool that produces specific recommendations for the next steps in gas development, and becomes the basic management information system for strategic analysis as more information and new options arise.

Once the basic GPM framework has been set up, it can be easily updated to accommodate better or more recent data. A continuously updated GPM can provide gas managers with a consistent framework to test quickly the effects of alternative investment decisions. As gas development proceeds, the GPM can also be expanded and made more flexible by incorporating disaggregated gas reserve data, for example, or better methods of demand forecasting.

In Part I we outlined an analytical framework that defines the important links between gas demand projections, gas supply options and the economic price path for gas over time, taking into account the value of the depletion premium. In Parts II and III we discussed the theory and practice of cost and benefit estimation as applied to natural gas projects. These provided the skeleton and components of a gas planning model. We now put them all together.

There are three general areas that a GPM should cover. First, it should develop a profile over time of the aggregate balance of demand and supply. This *sectoral context* is necessary in order to derive one or more scenarios for the economic price of gas over the relevant time-period. Second, it should identify, evaluate and rank alternative packages of gas-using projects and related infrastructure investments. This *project evaluation analysis* will be based on the gas price scenarios derived from the aggregate sectoral work. Third, *consistency checking and sensitivity analyses* should be undertaken in order to highlight any necessary revisions in the first two areas and to identify critical issues or information gaps that need to be filled before particular decisions on gas strategy can be taken. As is true in most pre-investment analysis, the judicious use of sensitivity testing in GPMs is not only essential to the credibility of the results, but also a valuable tool for defining the future work programme and the scope of later, more detailed feasibility studies.

Figure 12.1 illustrates these three parts of a GPM, the main components of each and the major relationships and connecting links among the components.[2] Although these are explained in more detail below, it is useful to begin with an overview of the process. The sectoral analysis (Stage I) begins with independent evaluations of aggregate gas demand (Box 1), as a function of the price and aggregate supply of gas (Box 2), on the basis of one or more alternative assumptions about

---

[2] The original framework of this model was set out in Julius, D., *Natural Gas Utilization Studies: Methodology and Application*, Energy Department Paper No. 24, World Bank, September 1985.

150 *The Economics of Natural Gas*

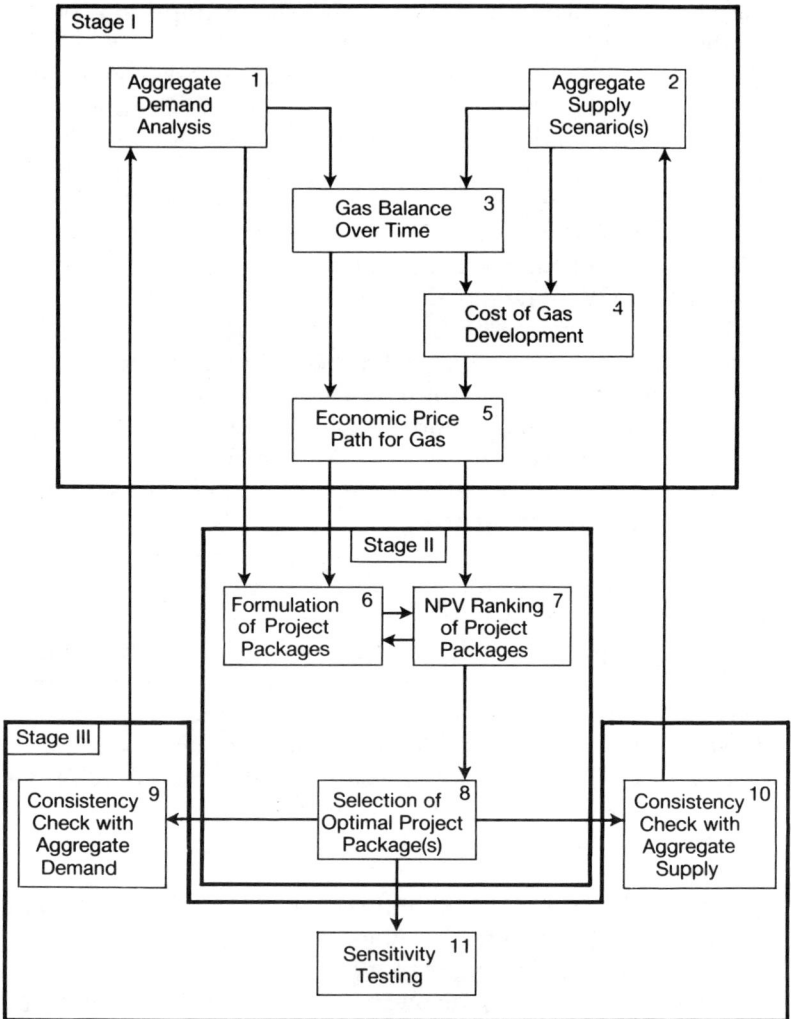

Figure 12.1: Components of a Gas Planning Model.

reserves. In Box 3 these are translated into time-dependent profiles of potential gas demand (at various prices) and supply (under various reserve scenarios). From this information and the costs associated with the investments for the aggregate supply scenario(s), the long-run marginal cost of gas can be calculated (Box 4). Using the framework described in Chapter 2, the economic price path(s) for gas is (are) then derived in Box 5.

Stage II constitutes the core of the GPM where alternative projects can be compared. First of all, projects are identified and grouped into tentative packages on the basis of their technical characteristics and economic complementarities (Box 6). In the largest gas-using sectors, such as electric power generation and the fertilizer industry, the aggregate demand analysis from Box 1 will already have identified a sequence of projects. To these will be added gas-using projects from other sectors and the export market, and net present value (NPV) calculations will be performed on the various trial packages (Box 7). If there is more than one gas price scenario, then correspondingly numerous 'optimal' project packages may be selected in Box 8.

In Stage III the gas consumption streams implied by each optimal package are checked against the corresponding aggregate demand and supply scenarios (Boxes 9 and 10). Any divergence should be traced through the steps in Stage I to see if there is a significant impact on the gas price path. If so, another iteration through Stage II will be needed. Once a consistent set of optimal project packages has been derived, sensitivity analyses should be carried out (Box 11) in order to test the robustness of the results and to identify the critical areas of uncertainty in the analysis.

In the following sections we discuss each of the GPM's components, paying particular attention to the demand analysis (Box 1) and the selection and ranking of projects (Boxes 6 and 7), where mistakes are most often made. As discussed in Chapter 2, for certain types of countries, some of the components shown in Figure 12.1 will be unimportant or can be handled by general estimates. In countries where the gas market is segmented (i.e. where particular reserves are linked to particular consuming centres, without interconnection), a separate GPM for each market will be needed. The option of interconnection at some future date can then be explicitly considered.

12.3 The Sectoral Context

(a) *Aggregate Demand Analysis (Box 1)*. Forecasting the potential future demand for a commodity that is not yet widely available is a difficult task, which is subject to considerable uncertainty. Furthermore, for a GPM both aggregate demand estimates and detailed information on the individual projects that will comprise the demand are needed. Such macro and micro analyses often require different types of expertise (e.g. economics and engineering) yet the two approaches must produce a consistent set of projections.

Before discussing the approach we recommend for demand analysis in a GPM, it is useful to describe the two approaches that are most often used. The first, and most common, is micro or 'bottom-up' demand estimation. This consists of identifying specific projects (e.g. a cement plant that could be converted to gas), and estimating and summing their gas requirements. While this approach appears to have the advantage of being clearly anchored to reality, it suffers from two important limitations. First, it is dependent for its success on a thorough knowledge of all the country's potential gas-using sectors, e.g. the industrial base, the electric power system and agriculture (as it affects fertilizer demand). Not enough gas planners will have the time or range of skills to obtain such knowledge unless the country is very small and underdeveloped. Secondly, using this approach, the estimation of levels of gas demand beyond the next five years or so is generally done very crudely. Yet it is the medium- and longer-term prospects for gas use that usually constitute the main focus of the GPM. Without an explicit grounding in the country's macroeconomic conditions and in the broader options for the overall energy sector (particularly for electricity), the projects generated by bottom-up extrapolation may miss the mark badly.

The second approach used in some studies of gas demand is macro or 'top-down' estimation, in which estimates for macroeconomic parameters, such as the GDP growth rate and the increase in the share of the industrial sector, are used to project gas demand. There are also two problems with this approach. First, in countries where gas is still a new fuel, the base figures from which the extrapolations are made are often too low. In the early years of gas development, consumption may double or triple annually since there are once-for-all opportunities for fuel switching, and consumption is generally constrained by supply. Macroeconomic parameters such as the growth rate of the country's industrial sector will be directly related to gas consumption only once the initial supply constraints and once-for-all market penetration investments are over. The second problem with exclusive reliance on the top-down approach is that the estimates produced will be only as good as the model from which they come. Most macroeconomic models are not sufficiently disaggregated to capture intra-sectoral changes (such as the pattern of fuel use in electricity generation) which may critically affect gas demand.

The bottom-up approach has clear attractions for near-term demand projections (say from one to five years). The top-down method can produce reliable projections for the long term (say beyond fifteen years), provided the projection period starts with a reasonable base level of consumption. Neither approach is likely to produce good figures for the medium-term period (say from five to fifteen years) – the period of greatest interest for most countries, given the lead times required for both infrastructure and gas-using project investments. It is in the medium term that the transition is made from a supply-constrained market (where consumption grows as quickly as supply expansion can take place) to a mature, demand-determined market (where most once-for-all opportunities have been exploited and consumption grows in line with the demand for the end products such as fertilizer and electricity that gas is used to produce). During the medium term, the availability of gas can provide the opportunity for structural changes in those end-use sectors themselves, for example, the choice between a gas- or a coal-based power development programme. This type of change should be a principal focus of the GPM.

To estimate gas demand for this critical medium-term period, a sectoral – rather than a micro or macro – analysis is needed. The GPM should analyse in depth the main gas-using sectors and their alternatives to gas in terms of both supply and investment. This time-consuming exercise should be limited to those sectors in which potential gas demand is very large, i.e. electric power, fertilizers and sometimes cement. Since these sectors provide the major opportunities for gas-using projects, an understanding of the determinants of their demand will aid both the estimation of aggregate gas demand (and thus the economic price path of gas) and the identification of individual project opportunities for gas use (Box 6 in Figure 12.1). The analytical treatment for each of the three main gas-using sectors is outlined below.

(*i*) *The Electric Power Sector.* In order to estimate the medium- and long-term role of gas in a country's power system, the GPM must generally rely upon the same tools that power planners use in projecting power demand and developing their least-cost expansion path for power investments. As illustrated in Chapter 8, the introduction of gas will not be a simple matter of substituting a planned hydro or coal-fired plant with an equal-sized gas-fired plant. Because gas-using equipment has lower capital costs (per megawatt of electricity produced), and therefore less pronounced economies of scale, a least-cost expansion programme based on gas may involve more frequent, smaller investment increments than one based on, say, coal. This ability to match demand and supply more closely may, in turn, imply that power based on gas will prove to be less costly than power based on coal even if the economic price of gas per mmbtu is higher than that of coal. For reasons such as these, a simple plant-by-plant comparison of gas generation versus alternative fuel-burning units will rarely be an adequate basis for projecting the power sector's demand for gas.

(*ii*) *The Fertilizer Sector.* Because of the dominant role of agriculture in the economies of many developing countries, and because of the continuing shift within the agricultural sector from traditional to modern farming methods, the domestic production of fertilizer often offers a large and growing market for natural gas. On the other hand, the NPV of gas-based fertilizer production may not be particularly high. For all but the largest developing countries, it is no longer economic to build diesel- or fuel oil-based plants to produce ammonia and urea rather than to import them: thus the value of gas in fertilizer production will often be lower than in other uses where it replaces fuel oil directly. Further, when the gas is used to produce fertilizer for export, rather than to replace imports, its value falls significantly. For these reasons, in order for a GPM to assess properly the amount of gas that would be used in the fertilizer market at various gas prices, it is necessary to develop an understanding of the country's present and future levels of demand for fertilizer.

This analysis will be considerably simpler than that needed for the power sector because economically-sized urea plants are quite large in relation to the market in most developing countries and there are few design issues that would significantly affect the amount of gas consumed. As discussed in Chapter 8, the principal relevant questions are those of project timing and the degree of export orientation of each plant over its lifetime. In the early years, the potential for switching existing plant from some other feedstock to gas should also be examined and compared with the option of replacement.

Once a schedule for investment in fertilizer plant to meet domestic needs has been drawn up, capacity utilization rates (for each plant for each year) should be calculated from the projections of fertilizer demand. This is important because economies of scale often dictate the need to build large facilities even where they may not reach full production levels for several years. Yet the capital investment required is so large, relative to that involved in other uses of similar quantities of natural gas, that capacity under-utilization can significantly lower a project's NPV.

(iii) The Cement Sector. For the purposes of a GPM, it is even more straightforward to project the demand of the cement industry than that of the fertilizer industry. Because it is seldom economic to export cement, investment in the industry will follow the growth of domestic demand, perhaps with an initial period of rapid expansion to substitute for imports. Plant sizes are often large in relation to the market, with many low-income developing countries needing only one or two new plants per decade. The choice of technology for a new plant will clearly favour the energy-efficient 'dry' process, and it will often be economic to convert or replace old 'wet' process

plants even if the economic price of gas is relatively low. The main economic issues in projecting the cement industry's demand for gas are the growth rate of the domestic market (and thus the scheduling of new plant) and the basis for measuring benefits. The latter will depend on the alternative, at the margin, to building a gas-based plant. As discussed in Chapter 6 above, if the country would otherwise find it economic to build a plant based on fuel oil or coal, then the benefit derived from gas will be based on the differential capital and operating costs of a plant using that alternative fuel. On the other hand, if the country's best alternative to building gas-based plant is to import cement, then the benefit will be derived from the projected import prices it would face.

(*b*) *Construction of the Demand Curve.* The demand analysis for the three sectors described above may need to be augmented in some cases by similar studies of other sectors in which significant gas use is expected to take place (e.g. LNG export, as discussed in Chapter 7) or in which it is likely to grow in a series of discrete steps rather than in a smooth build-up over time. Once this sectoral analysis has been completed for the largest users, the information must be combined to give a picture of the aggregate gas demand function. This can be done in five steps. The first step is to calculate the netback value of the gas consumed in each of the large gas-using projects outside the power sector. As explained in Chapter 6, the netback value of gas N is the total discounted NPV of the project with a zero cost of gas (in constant price monetary units) divided by the total discounted quantity of gas consumed (in volumetric units):

$$N = \frac{\sum_{t=1}^{T} [(B_t - I_t - O_t)/(1 + r)^t]}{\sum_{t=1}^{T} [Q_t/(1 + r)^t]} \qquad (12.1)$$

where B_t represents the project's benefits in year t, I_t the investment costs incurred in year t, O_t the operating and maintenance costs of the project in year t excluding the cost of gas, Q_t the consumption of gas by the project in year t, r the opportunity cost of capital, and T the time-horizon for the project.

The netback value thus represents the highest average price for gas that the project can pay such that it will just break even over its lifetime, relative to the best 'without-gas' alternative. As noted above, the alternative could be either a similar plant based on a different fuel or feedstock or the importation of the end product, whichever would be more economic in the absence of gas.

The second step consists of determining the power-related gas demand by postulating several trial price paths as discussed in Chapter 8. For this preliminary round a constant price can be assumed. Once the potential for converting existing plant to gas has been taken into account, the process can be simplified by seeking the optimal share of gas for a few widely spaced test years. This can be done on the basis of annuitized investment costs and fuel requirements which assume that for a major part of its life a plant is utilized at a fairly constant capacity factor. The result would resemble Figure 12.2.

The third step is to use the information about netbacks to graft the relevant demand for gas from other large projects on to the set of demand curves for the power sector. Suppose, for example, that two additional projects are considered: (a) a project with a netback of $4.00/mcf that uses 40 mmcf/d from 1990 onwards; and (b) a second project with a netback of $2.00/mcf to be commissioned in 1995 and which will

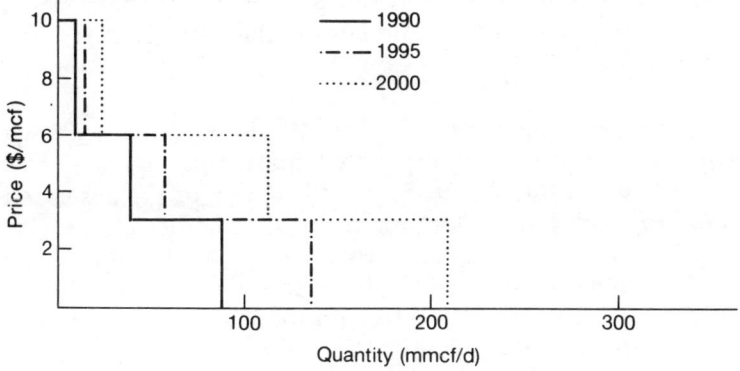

Figure 12.2: Demand for Gas by the Power Sector. 1990–2000.

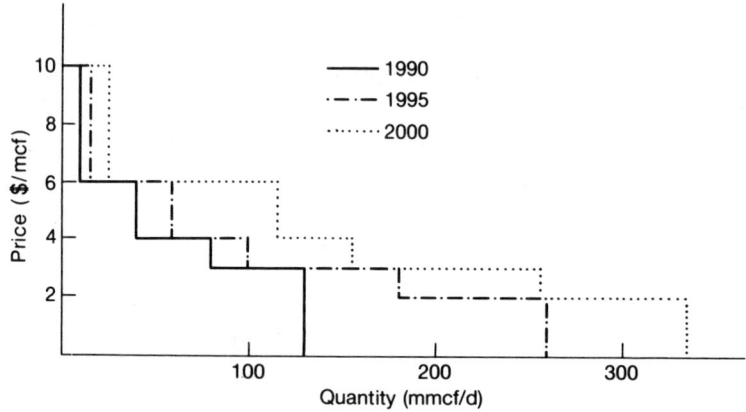

Figure 12.3: Demand for Gas by the Power and Other Major Sectors. 1990–2000.

consume 80 mmcf/d from start-up. Figure 12.3 shows how this information could be added to that on the power sector shown above.

The fourth step is to translate the price-versus-quantity graphs into quantity-versus-time graphs. This is done by connecting the points from the various demand curves over time at each price level P.

The fifth and final step is to add to this gas demand from the major sectors the demand from other users to arrive at a set of time-dependent aggregate demand curves. These incorporate the micro estimates of demand for the early years of the period, as well as the macro projections of total gas demand for the intermediate and later years. The result is shown in Figure 12.4.

(c) Aggregate Supply Scenario(s) (Box 2). The derivation of gas supply scenarios is generally more a matter of judgement than of firm information. Because the information on gas reserves is bound to change as exploration proceeds, and because a GPM is usually undertaken at an early stage in the development of gas, it is seldom worth while to invest much time or effort in fine-tuning estimates of reserves on the basis of current information. Rather, the views of geologists and other experts working in the country on future success ratios and potential

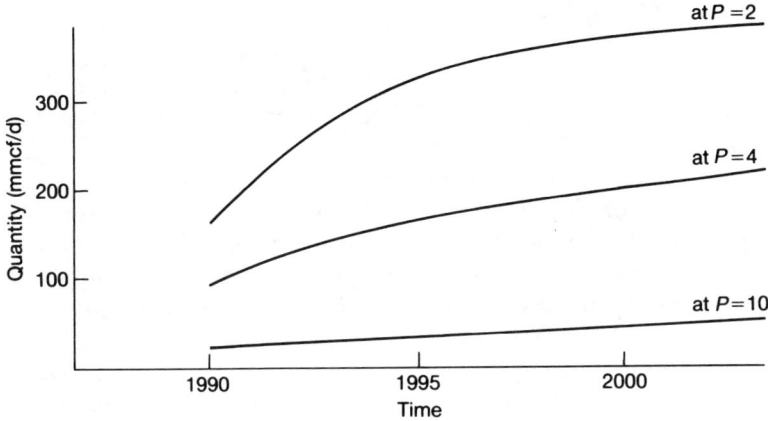

Figure 12.4: Aggregate Demand for Gas. 1990–2000.

discoveries that could be brought on stream in the period under consideration should be canvassed and assessed. It will often be appropriate to develop a set of two or three reserve figures which cover the probable range of estimates. For each reserve scenario, an estimate should be made of the total gas supply for each year of the period under study, on the basis of the existing infrastructure and the stage of discovery or development of the various fields. This will involve estimates of potential production from each major field or set of fields, taking into account field use and production decline rates as the fields reach maturity.

In addition to providing projections of gas supply that can be compared with the aggregate demand figures to identify periods of potential surplus and deficit, the aggregate supply analysis serves two other purposes. The first is to develop a tentative sequence of gas development projects by identifying the major supply options. These can form the basis of some of the project packages whose NPVs are calculated and compared in Stage II of the GPM. The second purpose of the supply analysis is to provide the investment framework for calculating the marginal cost of gas supply as an input into the economic price path for gas. The latter will be particularly important in gas-surplus countries where the depletion premium component of the price may be insignificant (see Chapter 2).

160   *The Economics of Natural Gas*

(*d*) *The Gas Balance Over Time (Box 3)*. The aggregate demand curves can now be matched with the maximum supply scenario(s) derived in Boxes 1 and 2. This will provide the first round estimate(s) of T^*, the point of economic depletion at which potential demand exceeds supply at the relevant price.[3] It is at this point that the informed judgement and close interaction of the economic and engineering experts on the GPM development team becomes important. It is generally possible for those who prepared the demand and supply estimates jointly to develop two or three iterations of the type described in Chapter 2, which will result in an optimal gas consumption path and thus an optimal T^* associated with each reserve scenario. Chapter 2 provides illustrations of several cases where this step will be self-evident (e.g. in a gas-short country), as well as a description of the circumstances under which it will be more difficult.

(*e*) *The Cost of Gas Development (Box 4)*. In Box 3, an optimal gas supply schedule over time is determined for each reserve scenario. Each will imply an investment programme for exploration and field development, and possibly for transmission and distribution. The costs of these are now estimated, using economic prices for all inputs. As set out in Chapter 4 above, the present value of the stream of investment and associated incremental operating costs can be divided by the discounted volume of gas consumed (as developed in Box 3) to yield the average incremental cost (AIC) of gas development for each reserve scenario.

There are several decisions involved in making the AIC calculation. The first is to determine which costs should be included in the general calculation and which should be allocated to particular users in Stage II of the GPM. In countries with diverse sources of gas and little existing use, it will generally be appropriate to calculate a well-head cost. Major gas-using facilities will often be set up close to the fields or the point of landfall, with little additional infrastructure cost. In other cases, where incremental supplies will be used mostly in

---

[3] There will be a different T^* for each reserve scenario.

distant urban and industrial areas or where they will feed a pipeline grid, it will be appropriate to calculate a city-gate cost. In general, the *AIC* should include all incremental system costs that cannot be clearly allocated to particular consumers or consumer groups. Allocable costs should be included in the gas-using project packages assembled in Box 6.

A second issue in the cost calculations is the degree of precision that is needed. The development of detailed cost estimates for a twenty-year investment programme is time consuming, and must be based on many, often arbitrary, assumptions. The empirical studies of eight developing countries discussed in Chapter 5 showed that the range of city-gate *AICs* was surprisingly small, considering the wide differences in the countries' circumstances. This implies that detailed cost estimates will not be necessary to yield an *AIC* with a level of confidence appropriate for the GPM. However, where there are major issues of gas project design (e.g. onshore versus offshore pipelines), a careful costing of each option will be important for the comparison of the NPVs for such alternative project packages in Box 7.

(f) The Economic Price Path for Gas (Box 5). The derivation of the gas price path is a straightforward application of the analysis discussed in Chapter 2. The depletion premium for each reserve scenario can be calculated directly from the point of economic depletion T^* developed in Box 3. This is added to the *AIC*, as determined in Box 4, to yield a gas price path over time for each reserve scenario.

12.4 Project Evaluation Analysis (Stage II)

In Stage I, the focus of the GPM is on the entire gas sector for the time-period under study. The objective of Stage II is to provide an early screening, or comparative testing, of major project candidates. It will generally include analyses of only the major projects under consideration, so that the project packages that are formulated and the NPVs that are calculated will usually represent only part of the expected investment in the sector. However, the analysis must be carefully designed to highlight all important investment choices.

(a) *Formulation of Project Packages (Box 6).* One of the most difficult and potentially time-consuming parts of a GPM is the formulation of trial packages of gas-using projects. Because of the complementarities involved in both investments (e.g. a transmission pipeline to an industrial complex) and products (e.g. methane for ammonia and ethane for ethylene), it is generally misleading to compare individual project alternatives. Furthermore, since gas is depletable and its price changes over time, questions of project sequence must be addressed explicitly. Although it is possible to generate and compare a large number of permutations of project packages, it is usually preferable to pre-select the most likely set of projects and then test basic inclusion/exclusion questions and compare different sequences of projects within each trial package. Project pre-selection should begin with the sectoral studies carried out in Box 1 for the major domestic users of natural gas. These will provide an initial set of projects sized and phased to serve the local market. For example, in the early years the package might include conversion projects for power stations and cement plants currently using diesel or fuel oil, followed by new gas-fired steam-generating plants and ammonia/urea facilities during the intermediate period, perhaps with additional power and cement plants over the longer term as demand grows. Once this base-case project package to serve the domestic market has been assembled, it is often useful to calculate the total 'project package net present value' (PPNPV) and to explore the potential for increasing the PPNPV through changes in the timing, sequence or basic design of the projects. The objective is to identify the best (i.e. highest-PPNPV) base-case package against which alternative packages can be compared.

To this base case various export-oriented projects can be added to see the effect of each on the total PPNPV. In addition, basic questions of gas system design can be explored in this stage (e.g. the size of a transmission pipeline to a new area).

Figure 12.5 shows a simplified example of project packaging. Case I is the simplest, in which an offshore pipeline is built to bring gas on shore, which then flows through an onshore pipeline to be used as fuel in an existing power station. Case II involves the addition of an LPG extraction

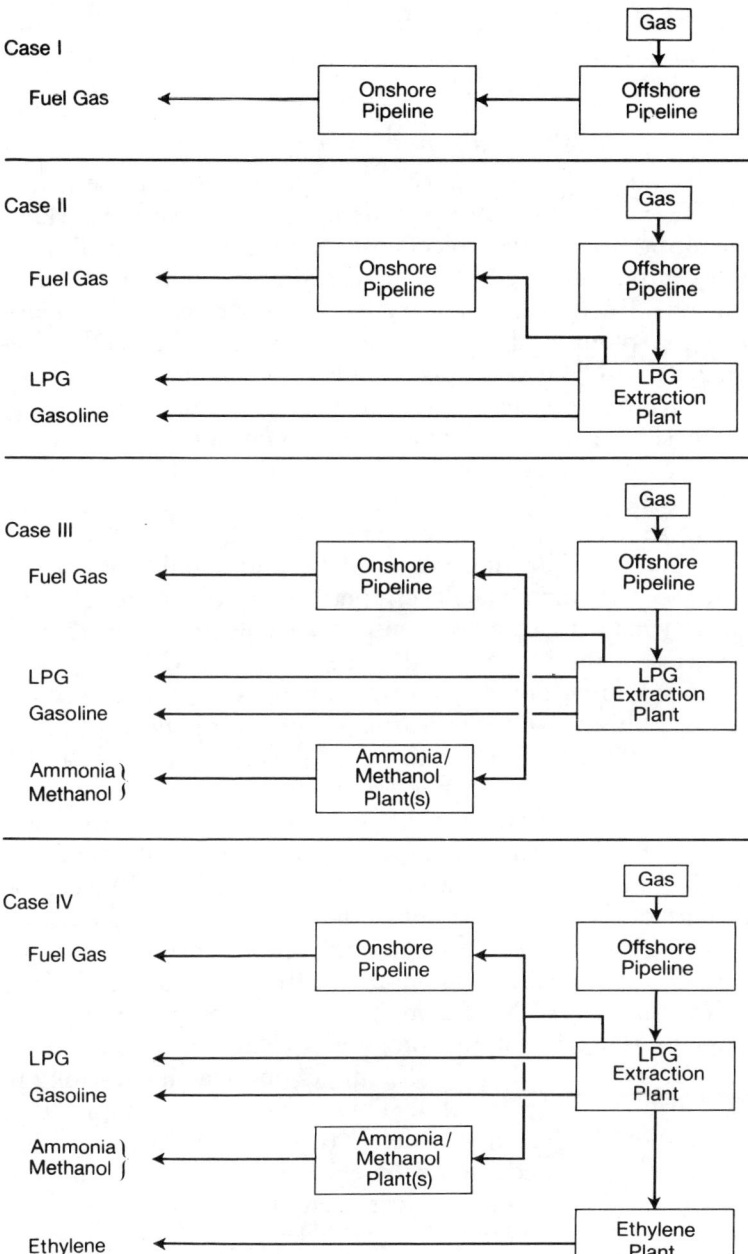

Figure 12.5: Construction of Sample Project Packages.

plant with its products entering the market as LPG and natural gasoline. In Case III, ammonia and methanol plants are added, and in Case IV, an ethylene plant is also included.

(b) NPV Ranking of Project Packages (Box 7). Although it is not conceptually difficult, the construction of data for use in the PPNPV ranking is often a time-consuming exercise. Many consulting firms have project cost information in their files on, for example, urea plants of a certain size built in various locations. Such costs can vary by a factor of two when comparing a plant built in a developed area with good infrastructure with the same plant built in a remote area with no existing infrastructure; it is important to tie them specifically to the site in question. Even then, such cost estimates are likely to vary by ±25–50 per cent, which should be borne in mind in later sensitivity testing. In addition to site-specific cost estimates, operating parameters such as start-up time and capacity utilization vary widely, and should be derived from those of similar facilities operating in the same country. For projects with a tradable output, such as urea or methanol, scenarios of the future trend in world prices will have to be developed or adopted from an outside source such as the World Bank or one of the major oil companies. If these products are produced for the export market, rather than for import replacement, then a broad analysis of the regional market for such products will be needed for sensitivity tests.

Once the necessary assumptions have been explicitly made and the data have been gathered, the various PPNPVs can be calculated, using any readily available computer software for spreadsheet analysis. A surprisingly common mistake, however, is to confuse economic and financial parameters. It is critical that the PPNPV comparison be made on strictly economic grounds, i.e. on the total net benefits that will accrue to the country rather than on the sum of net benefits accruing to each firm or project entity. This means, for example, that profits should be calculated before taxes rather than after, that the input costs used should be exclusive of any subsidies or price controls that may distort their market prices, that all investment costs should be included as the money is spent rather than when the loans raised to finance them will be repaid, that all costs should be on a constant price basis excluding inflation, etc. There are many good texts on such details of economic project evaluation that can be consulted for guidance.[4]

As the calculations proceed, it will often become clear that certain sets of projects always dominate others, or that certain sequences of projects always yield higher PPNPVs than others. For example, the production of ethylene derivatives may always increase the PPNPV, whereas the addition of propylene derivatives may decrease the total PPNPV. In this case, the more elaborate project packages – which would include, say, methanol as well as petrochemical production – should be reformulated, if necessary, to encompass ethylene derivatives but not propylene derivatives. The objective in the ranking exercise is to compare each larger project package with the *best* version of the previous package. This will generally require formulating some new project packages as the ranking results are generated.

(*c*) *Selection of Optimal Project Package(s) (Box 8).* For each gas price scenario (i.e. each gas reserve scenario from Box 2), the project package with the highest total PPNPV should be selected. If the same package is chosen under all price scenarios, then reserve uncertainty is not a constraint to gas utilization. If the first few projects (in time sequence) are the same in optimal packages for the various price scenarios, then developments for at least those projects can proceed without delay. In the unlikely event that the project packages for different reserve scenarios are completely different, with the same package ranking very high and very low on different reserve assumptions, then it may be best to step up gas exploration efforts and delay decisions on gas use until the reserve uncertainties can be reduced. In general, however, this exercise will identify some projects that clearly have a high investment priority (under all reserve scenarios), some that are sub-marginal (under all reserve scenarios) and some that become interesting only under the high reserve scenarios,

---

[4] See, for example, Squire, L. and van der Tak, H., *Economic Analysis of Projects*, Johns Hopkins University Press, Baltimore, 1976.

or whose feasibility is sensitive to parameters other than gas reserves. The sensitivity tests in Stage III should be used to segregate projects more clearly into these three categories.

12.5 Consistency Checking and Sensitivity Analysis (Stage III)

Pre-investment studies, such as a GPM, are by their nature characterized by substantial uncertainties. The further into the future they look, the greater the potential for significant error. As noted at the beginning of this chapter, two of the three primary objectives of a GPM are to define the critical areas of uncertainty that affect project choice and to recommend the types of analysis and data collection that should be undertaken in the more detailed feasibility studies for individual projects. These tasks require a rigorous and carefully designed sensitivity analysis, upon which the final credibility of all of the GPM's conclusions will rest.

(a) Consistency Check with Aggregate Demand and Supply (Boxes 9 and 10). Because of the logical connections between the gas demand and supply profiles, the economic price for gas, the selection of projects and total gas demand, it is necessary to check that the project packages selected in Box 8 remain consistent with the demand and supply profiles developed in Boxes 1 and 2. If, for example, a selected project package requires a faster build-up in gas supply than is possible under the relevant reserve assumption, then some reduction or rescheduling of projects within that package will have to be formulated and tested. If the selected package implies a lower rate of growth of demand than had been projected in Box 1, this may postpone T^* for a few years which, in turn, will lower the economic price of gas for the period before the new T^*, thereby requiring the calculation of a new set of PPNPVs. In most cases, however, given the range of uncertainty around the estimates in Boxes 1 and 2, minor inconsistencies will not result in large enough changes in the price path for gas to justify another round of PPNPV calculations.

(b) Sensitivity Analysis (Box 11). A well-designed sensitivity

A Practical Approach to Gas Planning   167

analysis should enable the GPM to provide firm recommendations on:

(a) A set and sequence (or order) of projects that are clearly attractive under most reasonable circumstances, and for which feasibility work should begin;
(b) A set of projects that are clearly unattractive under most reasonable circumstances, and for which no further studies should be undertaken at this time;
(c) A set of marginal projects whose feasibility will depend on certain identified parameters on which either more information is needed or more detailed analysis should be undertaken before or during feasibility studies.

The first step in designing the sensitivity analysis is to identify the set of physical and economic parameters whose magnitude and probability of occurrence are such that they could have a significant impact on the ranking of projects. These will often include world price projections for some project outputs (e.g. urea and LPG), assumptions about the growth of domestic demand for others (e.g. power and cement), variations around the chosen discount rate (where the packages include some capital-intensive projects), variations in the rates of capacity utilization (especially for export-oriented projects), etc. They should also include relevant future policy decisions where the linkages with the gas sector are strong, such as whether or not to upgrade a local refinery in order to reduce supplies of surplus fuel oil that would otherwise substitute for gas. Some of these uncertainties may have such a large effect on the ranking of projects that they should be directly built into the analysis of Stages I and II of the GPM. For example, major variations of reserve estimates should be incorporated into Box 2 and carried through the entire analysis. A major strategic question such as whether to supply gas to two areas of the country from the beginning or to start with one area and add the other later should be tested in the project package formulation and ranking of Boxes 6 and 7.

There is a danger that too many different sensitivity tests may obscure, rather than clarify, the ranking of project packages. There are various methods of minimizing this risk. It is often possible to group certain types of uncertainty into a single sensitivity test. For example, (a) delays in project implementation, (b) capacity under-utilization, or (c) a slower-than-expected build-up in local demand will all have the same effect of slowing the rate of gas consumption, thereby postponing T^* and lowering the price path of gas. A single sensitivity test on a lower gas price path could be used to test the effect of all three areas of uncertainty.

Some uncertainties offset one another so that, taken together, they have little net effect on project selection. For example, in estimating the cost of gas development in Box 4, the time-path of gas consumption that goes into the denominator of the *AIC* may be underestimated. However, this would imply that the schedule of investment used in the numerator of the *AIC* is also underestimated; so the net effect is probably small. It is often useful to designate one or a small number of base-case project packages from those selected in Box 8, and to apply the general sensitivity tests, such as using alternative discount rates, only to those cases. By using such base cases, by eliminating sensitivity tests of offsetting effects and by grouping uncertainties that result in similar effects, the number of sensitivity tests can be held to a reasonable level.

Table 12.1 presents an example of sensitivity test results. The first column represents the base-case test, T–1. The next four columns, T–2 to T–5, are different project packages using the same assumptions as the base case to calculate their PPNPVs. The next three columns, T–6 to T–8, are sensitivity tests on the base-case project package assuming: (a) coal replacement in power rather than fuel oil, T–6; (b) all incremental LPG is exported rather than half used domestically, T–7; and (c) a combination 'worst case' of these assumptions, T–8. Next come four policy alternatives, T–9 to T–12, which use the same project package as the base case. These are: (a) that a planned local refinery is not built, T–9; (b) that the pipeline route is mainly off shore rather than on shore, T–10; (c) that the pipeline is built in one stage rather than two, T–11; and (d) that the LPG projects are located in a different area, T–12. Finally, T–13 represents the 'do-nothing' option, assuming that only the existing gas development plans are carried out. The bottom line shows the PPNPVs, which indicate that T–9 is the preferred alternative.

Table 12.1: Sample Presentation of Sensitivity Test Results for Different Cases.

	T-1	T-2	T-3	T-4	T-5	T-6	T-7	T-8	T-9	T-10	T-11	T-12	T-13
Products:													
Fuel gas	Yes	Yes	Yes	Yes	Yes	Yes	Yes	Yes	Yes	Yes	Yes	Yes	Yes
LPG	Yes	No	No	Yes	Yes	Yes	Yes	Yes	Yes	Yes	Yes	Yes	Yes
Natural gasoline	Yes	No	Yes	Yes	Yes	Yes	Yes	Yes	Yes	Yes	Yes	Yes	Yes
Ammonia	Yes	No	Yes	Yes	Yes	Yes	Yes	Yes	Yes	Yes	Yes	Yes	Yes
Methanol	Yes	No	Yes	No	Yes	Yes	Yes	Yes	Yes	Yes	Yes	Yes	No
Ethylene derivatives	Yes	No	No	No	Yes	Yes	Yes	Yes	Yes	Yes	Yes	Yes	Yes
Propylene derivatives	No	No	No	No	Yes	No	No	No	No	No	No	No	No
Fuel gas value	Fuel oil	Fuel oil	Fuel oil	Fuel oil	Fuel oil	Coal	Fuel oil	Coal	Fuel oil	Fuel oil	Fuel oil	Fuel oil	Fuel oil
LPG market	Local/ export	Local/ export	Local/ export	Local/ export	Local/ export	Local/ export	100% export	100% export	Local/ export	Local/ export	Local/ export	Local/ export	Local/ export
Implementation of local refinery	Yes	Yes	Yes	Yes	Yes	Yes	Yes	Yes	No	Yes	Yes	Yes	Yes
Pipeline routing	On-shore	On-shore	On-shore	On-shore	On-shore	On-shore	On-shore	On-shore	On-shore	Off-shore	On-shore	On-shore	None
Pipeline phasing	2	2	2	2	2	2	2	2	2	2	1	2	0
Area of LPG location	A	A	A	A	A	A	A	A	A	A	A	B	A
Project package net present value (PPNPV)	23.4	15.7	16.4	22.6	23.2	19.9	21.2	18.3	27.7	23.0	22.1	23.0	7.1

Source: Adapted from Chem Systems Inc., *Master Plan Study for Natural Gas Utilization in Malaysia.*

12.6 Conclusions

The eleven-step network described in this chapter represents a compromise between theory and practicality. We do not propose, in the general case, the development of a simultaneous-equation model that can be solved for optimal demand and supply schedules, the gas price path and the set of optimal gas-producing and gas-consuming projects all at the same time. In many cases no such tool is needed, and its very sophistication tends to obscure the planning process.

The GPM proposed in this chapter is a fairly simple one, and could be further simplified (or expanded) by taking a different approach to individual 'boxes'. Depending on the stage of gas development reached by the country, and depending on the information already available, a suitable skeleton GPM can usually be built in three-to-six months by a two-to-three person team. As development proceeds and information on reserves, costs and the market accumulates, the skeleton GPM can easily be modified or expanded to remain up to date.

The real usefulness of the GPM will become apparent when a sudden opportunity arises; for example, a feasibility study in the power sector shows the next hydro project to be more costly than anticipated, or an international company approaches the government with an idea for a joint-venture fertilizer export project. Such options are normally complex to evaluate, but early commitments on gas prices and availability will be necessary to capture the market. With the GPM as a management tool to guide and speed its analysis, the gas company can respond in a confident and timely manner.

In addition to aiding in investment planning, a gas planning model can effectively break the 'chicken-and-egg' dilemma that too frequently stifles the rapid and economic growth of the gas sector in developing countries. It makes explicit to the policy-maker and the investor the links that connect the exploration and discovery of gas with its final use in the market-place. It makes it possible to evaluate the many uncertainties in the full chain of gas development. It traces the impact of a decision in one area (i.e. one box of Figure 12.1) on the choices available in other areas. It provides a simple mechanism for breaking the complexity of gas development questions into manageable bits without losing sight of the total picture.

In our view, the lack of such an integrated approach to taking decisions about gas development constitutes the biggest single obstacle to rapid progress in gas utilization in developing countries and to the design of effective regulatory systems in developed countries. As discussed in earlier chapters, the size of gas reserves is not a binding constraint. The cost of gas development compares favourably with alternative energy sources. The possible domestic uses for gas are numerous and growing. The benefits from such uses are high. They will be even higher when the next cycle of rising oil prices takes hold of the market. If the appropriate planning tools can be built in advance, then energy users and governments – in both developing and developed countries – will find themselves more resilient to the inevitable surprises in the world energy scene. They will be better placed to tackle the new energy and environmental challenges of the 1990s, with the expanded role for natural gas that these will involve.

Index

Abu Dhabi, 32, 35, 37, 73, 75–7 *passim*
Afghanistan, 38
Africa, 33, 36, 37, 39, 40, 42, 61, 76, 77
   see also individual countries
agriculture, 100, 152, 154
air conditioning, 106, 110
Alaska, 73, 77
Algeria, 34–7 *passim*, 39, 73–7 *passim*, 102, 110
ammonia, 66, 70, 91, 95, 97, 154, 162–4 *passim*, 169
Angola, 39, 76
Argentina, 37–9 *passim*, 76, 102
argon, 30
Arzew, 73
Asia, 33, 36, 38, 40, 42, 76, 89, 92 *see also individual countries*
associated gas, 3, 12, 30, 31, 34, 34n7, 38, 49, 60, 70, 79, 87, 120n2, 146, 147
Australasia, 42
Australia, 38, 77

Bahrain, 110
Bakhrabad, 109, 128, 129, 132, 133
Bangladesh, 38, 76, 102, 105, 109, 125–44
Beani Bazar, 132
Belgium, 36
benefits, 1, 3, 5, 47, 52, 65–89, 109–11, 156, 164, 171
boil-off, 81, 81n5
Bolivia, 37, 38, 76, 105, 110
Bonny project, 39
Brazil, 38, 76, 102, 105
Brunei, 35, 36, 38, 74–7 *passim*
Burma, 38, 76
Burmah Gas Transport Ltd, 74
burn-off, 84
butane, 21, 30, 31, 99
buy-back clauses, 121

Cameroon, 39, 56, 58, 59, 76
Canada, 34, 35, 37, 77

capacity utilization, 69, 155, 157, 164, 167, 168
carbon dioxide, 23, 30, 31, 79
Caribbean, 37
Cedigaz, 41
cement, 19, 68, 95, 96, 152, 154–6, 162
Chem Systems, vii
chemicals industry, 95, 98–9
Chile, 38
China, 97, 98
coal, 1, 9–11 *passim*, 15, 16, 20, 21, 25, 26, 95–7 *passim*, 101, 102, 121, 145, 146, 148, 154, 156, 168
Colombia, 38
combined-cycle units, 93
commitment, government, 57
compression, 21, 24, 49, 56, 57, 60, 62, 79
condensates, 22, 129, 132
conservation, 27, 75
consumption, 15, 19–21 *passim*, 26, 29–30, 32, 34, 35, 37–44 *passim*, 48, 102–11 *passim*, 117, 122, 125, 134, 139, 151, 153, 160, 168
contracts, 74, 77, 115, 120–4, 120n2
conversion, 20, 41, 68, 92, 93, 96, 97, 122, 157, 162
cooking, 3, 10, 61, 65, 101, 102, 104, 106, 108, 142
cooling, 42, 110
costs, 1–3 *passim*, 5, 6, 10, 12–14, 17–27 *passim*, 34, 44, 47–62, 66, 68, 78, 80–4, 87, 88, 92–3, 96, 97, 102, 105–7, 118, 120–2 *passim*, 125, 154, 156, 160–1, 164–5
   AIC/marginal, 12, 13, 15, 52–3, 55–62, 68, 69, 71, 109, 122–3, 127–40, 151, 159–61, 168
   development, 18–22 *passim*, 47–53, 55–62, 66, 160–1, 168, 171
   distribution, 2, 51, 56, 103–10 *passim*, 125, 134–7, 146
   exploration, 48–9, 52, 55–7, 59, 60, 62, 87, 88, 109, 125, 127–8, 141, 142

joint, 50–1, 60, 61
liquefaction, 77–82, 87
opportunity, 9, 15, 25, 139, 157
production, 10, 12, 13, 15, 17, 19, 22, 25–7 passim, 34, 44, 49–50, 55–62, 77–83, 87, 88, 109, 118, 125, 129–33, 140–2
transmission, 2, 10, 24, 25, 27, 50–1, 55–7 passim, 59, 60, 109, 125, 133–4
transport, 1, 14, 44, 71, 81, 84, 87, 97
CPEs, 33, 41, 43 see also individual countries and Europe, Eastern

demand, 2–4, 9–11, 13–16, 18–25 passim, 28–44, 51, 57–60 passim, 67, 69, 69n3, 88–9, 93, 95, 101–4 passim, 118, 122, 126–7, 129, 133, 134, 140, 145–60, 166–8 passim, 170
depletion, 2, 4, 9, 15, 16–18, 24–7, 61–2, 160, 161
  premium, 15, 16, 18, 20, 25–8, 118, 125, 139–42 passim, 145, 149, 159, 161
developing countries, 2–4 passim, 14–15, 20–30, 32–5 passim, 37, 39–44 passim, 55–62, 66, 70, 73, 74, 76, 89, 91, 98–103 passim, 105, 108, 110, 111, 125–44, 146, 154–5, 161, 170–1 see also individual headings
development, 2–6 passim, 9, 10, 18, 19, 21–3 passim, 33, 34, 47–62, 66, 119, 146, 148, 160–1, 168, 170–1 see also costs; planning
diesel fuel, 10, 20, 95, 97
discount rate, 18, 22, 26–7, 51, 58, 65, 83, 85, 86, 93, 109, 111, 139, 140, 167
diseconomies of scale, 86, 87
distribution, 2, 14, 48, 51, 56, 66, 74, 82, 101–7, 109–11 passim, 121, 125, 133–7, 146, 160 see also costs
'dry gas', 21, 22, 24, 25, 27, 61, 133

economies of scale, 12, 19, 51–3, 57–8, 60, 61, 71, 75, 79, 81, 121, 154, 155
efficiency, 100, 108, 117, 118
Egypt, 39, 56, 58, 76, 102, 108
El Paso project, 39
electricity, 1, 2, 4, 14, 15, 19, 65, 91–5, 145, 153 see also power generation
environmental factors, 1, 42, 74, 76, 97, 101
equity, social, 117–18, 142

ethane, 30, 31, 95, 99, 100, 162
Ethiopia, 39
ethylene, 95, 100, 162–5 passim, 169
Europe, Eastern, 33, 35, 36, 40, 42
  Western, 33, 36–40 passim, 38n10, 42, 43, 74–7 passim, 81, 89, 102, 119 see also individual countries
exploration, 25, 28, 32, 48–9, 52, 55–7, 59, 60, 62, 109, 115, 119, 120, 125, 127–8, 158–60, 165 see also costs
exports, 9, 11, 14, 21, 22, 31, 32, 35–9 passim, 38n10, 44, 66, 70–1, 73–89, 121, 146, 147, 155, 156, 168

FAO, 98
Far East, 37
feedstock, 3, 10, 14, 41, 68, 70, 91, 95–9, 155
fertilizers, 1, 10, 11, 14, 16, 17, 19, 38, 41, 65, 68–70, 95–8 passim, 126, 146, 147, 151, 153–5 passim, 162, 163, 170
flaring, 2, 9, 32, 34, 35, 73–4, 146, 147
forecasting, 56, 57, 118, 120, 123, 126–7, 145, 148–71 passim
France, 36, 40, 73, 105, 107
fuelwood, 101

gas-short countries, 10, 19, 57, 66, 118, 119
gas-surplus countries, 10, 18, 57, 61, 73–4, 89, 118, 119, 159
gasolines, 49, 50, 164, 169
gathering systems, 34n7, 49, 59, 60
Ghana, 39
guarantees, 120, 121

Habiganj, 129–33 passim
heating, residential, 3, 14, 42, 43, 61, 65, 101, 102, 104–7 passim, 110, 111, 142
helium, 30
Hotelling, Harold, 17–18
hydropower, 20, 66, 93, 145, 147, 148, 170
hydrocarbons, 30, 80n2, 96
hydrogen, 30
  sulphide, 30, 61, 79, 96

IGAT I and II, 38, 38n10
imports, 11, 14, 21, 32, 35–8 passim, 70, 74, 96, 146, 154–7 passim
income, redistribution of, 117
India, 38, 56, 58, 59, 102
Indonesia, 35–8, 74–7 passim, 148

Index 175

inflation, 51, 65, 124, 165
infrastructure, 2–4 passim, 18, 21, 48, 49, 83, 86, 87, 91, 99, 102, 103, 122, 147, 149, 153
institutional strength, 57
International Gas Union, 33
interruptible service, 3, 121, 122, 141
investment, 2–4 passim, 12, 19, 24, 48, 50, 51, 53, 62, 68, 78, 88, 91, 95, 102, 120–2 passim, 133, 142, 145, 146, 149, 153–5 passim, 157, 159–61 passim, 164, 165, 168, 170
Iran, 29, 32, 34, 35, 38, 38n10, 75, 76, 101, 102, 105
Iraq, 35
Italy, 36, 37n8, 39, 40, 105
Ivory Coast, 39

Japan, 33, 36–8 passim, 40, 42, 73–7 passim, 81, 89, 96, 102
Japan Indonesia LNG Company, 74
Jensen Associates, vii
Kailashtila, 129, 132
Kennedy & Dunkin, vii
kerosene, 10, 101, 108, 146
Khuzestan, 38
Kuwait, 38

Latin America, 33, 36–8 passim, 40, 76, 109, 110 see also individual countries
lean gas, 30, 49, 50, 61, 80, 80n2, 132
Libya, 35, 37n8, 39, 74–6 passim
lignite, 147, 148
liquefaction, 22–3, 30, 31, 49, 71, 74, 75, 78–82, 84–7 passim, see also costs
liquids, 22–3, 27, 49, 61
LNG, 11, 14, 35–40 passim, 44, 61, 70–1, 73–89, 147, 156, 163, 164
load factors, 51, 121, 122, 126–7, 133, 141
Louisiana, 73
LPG, 10, 22, 30, 49, 50, 101, 108, 146, 162, 164, 168, 169

Malaysia, 36–8, 75–7 passim
management, 18, 48, 148
markets, 1, 6, 11, 14, 18, 31, 34, 41–4 passim, 61, 62, 91–111, 115, 116, 122, 151
metering, 49, 109, 122, 123, 140, 141
methane, 30, 31, 162
methanol, 10, 41, 91, 95, 98–9, 163, 164, 169

Mexico, 37–9 passim, 76, 102, 105, 147
Middle East, 32, 33, 36–8 passim, 40, 42, 76, 89, 147 see also individual countries
Miskar, 59
Mnazi Bay, 60
Mobil Oil Indonesia, 74
monopolies, 14, 67, 115
monopsonies, 15, 115
Morocco, 39, 56, 58, 59, 110

naphtha, 70, 95, 96, 98, 102
Netherlands, 35, 102
Nigeria, 29, 34n7, 35, 37, 39, 56–8 passim, 60–1, 75, 76, 105, 148
Nippon Steel Corporation, 74
nitrogen, 30, 31, 61, 79, 97–8
non-associated gas, 4, 12, 22, 34, 34n7, 38, 39, 61, 133, 146
North America, 29, 33, 36, 40, 42, 102 see also individual countries
Norway, 35, 37, 76

Oceania, 33, 36, 40
OECD, 33, 41, 97, 102
oil, 2, 4, 6, 9–11 passim, 13, 14, 19–21 passim, 30, 32, 34, 49, 50, 56, 61, 68n2, 70, 92, 95–7, 111, 119, 120n2, 121, 124, 145–7, 154–6 passim, 167, 168 see also prices
  companies, 32, 127–8
OPEC, 10, 29, 34, 35, 70
Osaka Gas, 74
ownership, 1n1, 74, 115

Pakistan, 38, 39, 43, 61, 101–3 passim, , 105, 110
peaking, 10, 20, 106, 121, 126–7, 136, 140
peak shaving, 73n1, 78, 93
pentane, 30, 31
Pertamina, 74
Peru, 38, 110
Petrobangla, 125, 128
petrochemicals, 1, 14, 69, 70, 91, 95–6, 99–100, 147, 165
pipelines, 1n1, 2, 13, 14, 21, 22, 24, 31, 35, 36, 38, 39, 51, 57, 60, 62, 74–6 passim, 85, 89, 105, 107, 125, 134, 136, 137, 161–3 passim, 168, 169
planning, 4–6 passim, 18–19, 71, 92n2, 145–71 see also forecasting
plastics, 100
political factors, 14–15, 39

176 *The Economics of Natural Gas*

pollution, 74, 77, 96, 101
polymers, 100
polyolefins, 99–100
population desnity, 42, 51, 77, 104, 105, 107, 110–11, 147
ports, 78, 81, 82
power generation, 1, 4, 10, 11, 14, 19–23 *passim*, 25, 39, 41–3, 51, 65, 66, 77, 91–5, 126, 141, 145–7, 151–4 *passim*, 157–8, 162, 168, 170
prices, 1–3 *passim*, 5, 6, 9, 10, 13–28 *passim*, 39, 44, 47, 57, 65, 69, 74, 76, 77, 83–9 *passim*, 93, 95, 105, 115–24, 140–4, 145, 149–51 *passim*, 154, 156, 157, 161, 164–8 *passim*, 170
  coal, 25, 26
  fertilizer, 98
  oil, vii, 6, 19, 21, 22, 39, 75, 77, 89, 97, 171
pricing, 1, 3–5 *passim*, 9, 26, 52, 67, 68n2, 69, 71, 77, 115–44
  objectives, 5, 116–18
processing, 22–3, 30, 61, 74, 96, 99, 129
production, 12–14, 20–2 *passim*, 24, 25, 29, 31, 32, 34–6, 48–50, 55–62 *passim*, 78, 79, 109, 118, 125, 129–33, 139, 140, 159 *see also* costs
profit/profitability, 17, 50, 69, 71, 121, 164
project selection, 4, 9, 19, 26, 67, 148–51 *passim*, 161–6
  sequence, 18, 19, 151, 159, 162, 165, 167
  size, 19
propane, 21, 30, 31, 99, 100
propylene, 95, 100, 165
PT Arun Natural Gas Liquefaction Company, 74

Qatar, 29, 38, 76

Rashidpur, 132
'raw gas', 21, 22, 27, 30–1, 61, 96, 99, 132
refrigerators, 102
regasification, 65, 71, 78, 79, 82–7 *passim*
regulation, 1, 3, 5, 14, 115, 171
reinjection, 30, 34, 35, 38, 49
rent resource, 3, 6, 15, 25, 115, 121, 140, 142
reserves, 6, 9, 10, 12, 13, 18, 20, 24–6, 28–34, 37–9 *passim*, 44, 48–9, 56, 59–62 *passim*, 74, 76, 78, 79, 89, 126, 128, 139, 140, 146, 148, 150, 158–61, 165–7 *passim*, 171

risk, 27, 48, 52, 88–9
royalties, 15

safety, 81, 81n3
Saudi Arabia, 35, 38
seasonality, 106, 121, 122, 140
sensitivity testing, 28, 65, 108–11, 149–51 *passim*, 164, 166–9
separation, 22, 27, 30, 49, 57, 61, 78, 79, 132
Shell, viii, 127
shipping, 74, 78, 79, 81, 85, 87
shortages, 19, 101, 127
Sofregaz, vii, 137
Somalia, 39
Songo-Songo, 60
South Korea, 38, 74, 76, 89, 103
steel, 11, 15
stocks, 15, 16, 18, 31, 59 *see also* reserves
storage, 78, 79, 82, 84, 85, 129, 147
subsidies, 116–19 *passim*, 124, 142, 164
substitutability, fuel, 10, 11, 14, 15, 25, 41, 68–9, 108, 111, 125, 147, 153, 168
Sudan, 39
supply, 3, 4, 6, 9, 12–13, 15, 16, 18, 19, 21–4, 28–44, 47–53, 57, 75, 88, 91, 95, 109, 118, 129, 140, 149–51 *passim*, 153, 158–60, 166, 170
'surplus-window' countries, 19–28 *passim*, 57, 118
Sylhet, 129

tankers, 36, 78, 81–2, 81n3,4, 83
Tanzania, 39, 56, 58–61 *passim*
tariffs, 2, 4, 5, 28, 115–17 *passim*, 121–4, 140–4, 145
taxation, 2, 3, 119, 140, 142, 143
technology, 31, 42, 99, 155
telecommunications, 57, 134n4
Thailand, 38, 56–9 *passim*, 61, 76, 148
Titas system, 125–44
town gas, 74, 102, 108
trade, 35–9, 73–89
  spot, 78
trans-Mediterranean project, 39
transmission, 10, 13, 24, 48, 50–1, 55–7 *passim*, 59, 60, 103, 109, 125, 133–4, 160 *see also* costs
transport, 14, 74, 81, 87 *see also* costs
Trinidad and Tobago, 38
Trunkline project, 39
Tunisia, 39, 56, 58–61 *passim*, 102, 105, 108, 110
Turkey, 103, 105, 110

Index 177

UAE, 35, 76
UNIDO, 98
United Kingdom, 36, 40, 73, 102, 105
United States, 1n1, 34, 35, 37–9 *passim*,
   40n11, 42, 43, 50, 73, 75, 76, 81, 89,
   102, 119
urea, 9, 66, 95, 96, 146, 147, 154, 155,
   164
use, of gas, 1–6 *passim*, 9, 10, 14, 15, 25,
   25n3, 39, 41–3, 47, 66–71, 73–89,
   91–111 *see also* consumption;
   feedstock
  commercial, 39, 42–3, 51, 66, 102,
     107–8, 110, 136, 137, 140, 143
  industrial, 10, 11, 20–2 *passim*, 39, 41,
     43, 66, 91, 95–100, 126, 136, 137,
     140, 141, 143, 146, 152–6 *passim*,
  residential, 3, 14, 39, 42–3, 51, 61, 66,
     101–7, 110, 136, 137, 140–3 *passim*
USSR, 32, 35, 37, 38, 38n10, 39–40, 42,
   43, 77

value, of gas, 9–11 *passim*, 16, 17, 25, 43,
   44, 50, 56, 84–9, 91–111, 115, 120,
   121, 140, 151, 154–5, 159, 161
  netback, 68–71 *passim*, 83–9, 93–100
    passim, 103–4, 108, 110–11, 156–7
  NPV, 87–9 *passim*, 110, 111, 150, 151,
    154, 159
  PPNPV, 162, 164–6 *passim*, 168, 169
  replacement, 118
vaporization, 81, 81n5, 83, 83n6, 87
Venezuala, 34, 38, 102
viability, financial, 117, 118, 142, 144

West Germany, 36, 102
World Bank, vii, viii, 5, 55, 65, 164

yields, 58–9